PROTEAN SUPPLY CHAINS

PROTEAN SUPPLY CHAINS
Ten Dynamics of Supply and Demand Alignment

JAMES A. COOKE

Published by John Wiley & Sons, Inc., Hoboken, New Jersey
Published simultaneously in Canada

For general information on our other products and services or for technical support, please contact our Customer Care Department within the United States at (800) 762-2974, outside the United States at (317) 572-3993 or fax (317) 572-4002.

Wiley also publishes its books in a variety of electronic formats. Some content that appears in print may not be available in electronic formats. For more information about Wiley products, visit our web site at www.wiley.com.

Library of Congress Cataloging-in-Publication Data is available.

Printed in the United States of America

ISBN: 9781118759660

10 9 8 7 6 5 4 3 2 1

To Lee

CONTENTS

PREFACE

Six years ago, I was asked to become editor of a new magazine aimed at providing thought leadership for the global supply chain. The magazine, Council of Supply Chain Management Professionals' (CSCMP) *Supply Chain Quarterly*, was a joint venture between the professional organization CSCMP and Agile Business Media, which publishes the North American logistics magazine, *DC Velocity*.

As editor of the magazine, I have written about the transformational supply chain practices of leading companies, reported on business trends sweeping the globe, and had the opportunity to edit the writings from some of the finest thinkers in the field. My work on the *Supply Chain Quarterly* provided the basis for many of the ideas explored in this book. In my talks with supply chain chiefs, management consultants, industry analysts, and academics, it occurred to me that businesses across a number of industrial sectors would have to move beyond agility and flexibility in their operations to master the volatile business environment in the global economy. They would have to develop "protean" supply chains capable of quick adaptation to unfolding events to ensure the continued profitability of their enterprises. The why, the what, and the how of protean supply chains are the subject of this book.

My thinking about protean supply chains has evolved over the course of editing *Supply Chain Quarterly*. The core ideas outlined in this book have appeared in a much briefer form in the column that

I write every month on supply trends and developments for a monthly electronic newsletter, *Supply Chain Quarterly Executive Insight*. Writing this book gave me the opportunity to delve into those ideas and to provide more depth and detail. It also gave me the opportunity to provide some context for people who do not work in the field of supply chains. In writing the column, I assume that my readers work in supply chains and understand the references, but in this book, I assumed the opposite was the case. One of my hopes for this book is to open the eyes of the general reading public to what is taking place in supply chains. Supply chain synchronization plays a major factor in the current high rate of joblessness affecting the United States. The development of protean supply chains will have an even greater impact on the United States and on world economy and will ultimately change how many people earn their livelihoods.

Although to my knowledge no company has a true protean supply chain, many companies are hard at work putting in place many of the elements of such a supply chain. Some of those companies are mentioned in this book. Unfortunately, most big companies do not like to discuss what they are doing, as they believe that their supply chain provides a competitive advantage in the marketplace. In a few instances, I mention the names of these companies, as their executives have been willing to discuss what they have done in articles in *Supply Chain Quarterly* and in *DC Velocity* magazine.

Although I quote many individuals in this book, I should note that the opinions on supply chain trends are mine. Some will disagree with my views; in fact, many will dispute my opinion on sustainability. Others will say that companies have tried to be agile in their supply chains for the past decade. The difference, in my opinion, is the pace of change. As companies look to match demand with supply, they will need to change manufacturing, distribution, and procurement operations frequently and easily. That is the essence of being protean.

JAMES A. COOKE

ACKNOWLEDGMENTS

After working on this project, I have come to the conclusion that it takes a village to write a book. In the year and half I spent on this project, I was fortunate to have help from many individuals who were kind enough to answer questions and, in some cases, point me in the right direction. There is a long list of folks to thank for their time and help.

Let me start with the academic community. During my research, I received help from many professors who both conduct research and teach supply chain practices and information technology in university classrooms. In particular, a number of professors at the Massachusetts Institute of Technology assisted me in my research. Edgar Blanco, Jonathan Byrnes, Chris Caplice, Yossi Sheffi, and David Simchi-Levi, all at MIT, patiently answered my questions. Other professors who offered input for this book are Larry Snyder at Lehigh University and Sean Williams at Boston University.

Gartner Research is the leading research firm covering business software and supply chains. Several Gartner analysts were willing to share their insights on software developments impacting supply chains. They include Stan Aranow, Dwight Klappich, Tim Payne, Kevin Sterneckert, and Noha Tohamy. Another analyst who provided assistance was Steve Banker of the ARC Advisory Group, with whom I worked on a research study about omnichannel issues in distribution that is cited in this book. I kicked around many of the ideas about the macroeconomic trends in this book with economist Chris Christopher of IHS Global, who writes a column for the *Supply Chain Quarterly*.

Because their work involves solving problems for companies, business management consultants are very close to what is happening in today's supply chains. A number of top consultants made themselves available to answer my questions. The folks at Tompkins Associates were generous with their time. The head of the firm, Jim Tompkins, as well as Gene Tyndall and Michael Zakkour, provided valuable input. Justin Rose and Claudio Knizek of Boston Consulting Group offered assistance, as did Kevin O'Laughlin and Rob Barrett from KPMG. Other consultants who contributed information are Ron Ash of Accenture, Paul Brody of IBM, Marilyn Craig of Insight Voices, Tom Craig of LTD Management, Sean Culey of Aligned Integration Ltd., Goetz Erhardt of Accenture, Frode Huse Gjendem of Accenture, John Haggerty of IBM, Helgi Thor Leja of Fortna, Rob O'Byrne of the Logistics Bureau, Martin O'Grady of Outsource Management Services, John Sewell of Crimson & Co., Daniel Swan of McKinsey, and Rob van Doesburg of Capgemini. For more information about control towers, which are discussed in this book, I recommend the white papers on this topic by Rob van Doesburg.

Many software vendors provide the applications that will enable the strategies outlined in this book. The following individuals from software firms offered their help: Hemant Bhave of Barloworld, Greg Brady of One Network Enterprises, Don Hicks of Llamasoft, Justin Honaman of Teradata, Howard Jensen of Viewlocity, John Lash of Terra Technology, Trevor Miles of Kinaxis, Kirk Monroe of Kinaxis, Kelly Thomas of JDA, Michael Watson of Opex Analytics, and Dennis Weir of Barloworld. I am especially indebted to D'Anne Hotchkiss at Teradata, who went beyond the call of duty to provide material from her company's conference. Terry Wohlers, one of the foremost experts on 3-D printing, provided background on that emerging technology, as did Sandeep Rana of AMT. Marcel Goldstein of CSC was helpful, as was Gordon Fuller of CSC. By the way, CSC has issued a few white papers on 3-D printing that I recommend as some of most detailed writing on this subject.

The following also provided input for this book: Charles Bogoian of Kenai Sports, John Gattorna, Jeff LeRoy of P&G, Harry Moser of the Reshoring Initiative, company owner Justin Norman, and inventory software expert Shaun Snapp. Others who provided assistance were Dexter Galvin of the Carbon Disclosure Project, Haley Garner with Eyefortransport, and Jamie Plotnek of the Carbon Trust. I would like to especially thank Matt Davis for his help as well on the topic of segmentation. I also have to give a special thanks to Peter Surtees and Alan Waller.

In addition to having the opportunity to learn from some of the best practitioners, academics, and consultants working in supply chains, I have been lucky to also work with some of the finest business journalists around. The quality of the *Supply Chain Quarterly*'s content owes much to the editorial skill and acumen of my coworkers, Toby Gooley and Susan Lacefield. Since I also write for *DC Velocity* and write a monthly column on technology trends for that magazine, I have also had the opportunity to work and learn from its editors: Karen Bachrach, Peter Bradley, David Maloney, Mark Solomon, and Jeff Thacker. This book draws in part on some of the industry coverage that has appeared in both publications. I also thank Mitch MacDonald, who oversees the editorial direction of both publications, for the good fortune to be involved with the *Supply Chain Quarterly*.

Finally, I'd like to thank Mini Villamayor, who copyedited this book, and my editor at Wiley, Susanne Steitz-Filler. And if I left anybody out, I offer my sincerest apologies.

J.A.C.

CHAPTER 1

SUPPLY CHAIN SYNCHRONIZATION

Since the Great Recession, the United States remains stuck in a jobless recovery. The same holds true for many other parts of the world. The U.S. economy is witnessing a profit recovery. Stocks have soared on Wall Street due to strong corporate profits, while unemployment in the United States has stayed stubbornly high above traditional norms. Although U.S. unemployment had dropped from its peak of 10% in October 2009, the Bureau of Labor Statistics still had the unemployment rate around 7% in late 2013. And it is not just the United States that is wrestling with joblessness. Unemployment remains high in other parts of the world as well. The 17-member Eurozone had unemployment rates above 12% in 2013.

This condition has been dubbed "the New Normal economy," a term coined by William Gross, the founder of the global investment firm Pacific Investment Management (PIMCO), back in 2009. Gross used the term to describe a lackluster economy he predicted would occur for a decade, an economy that would witness high unemployment and a reduced standard of living for Americans as well as most other citizens around the globe.

Protean Supply Chains: Ten Dynamics of Supply and Demand Alignment, First Edition. James A. Cooke. © 2014 John Wiley & Sons, Inc. Published 2014 by John Wiley & Sons, Inc.

One of the reasons behind the continued high employment rate, ignored by both economists and the media pundits, is the impact of supply chains on business efficiency that has resulted in strong corporate earnings. Companies have moved closer to achieving supply chain synchronization, although it should be noted they still have a long way to go before reaching perfection. But even a modest movement in that direction is having a strong economic effect. In theory, synchronization means that companies would make only the exact amount of goods necessary to meet actual consumer demand. Synchronization means that there is no excess inventory throughout a supply chain spanning continents. Synchronization of supply with demand leads to increased output without the need for additional labor.

Aligning production with consumption has been a business goal for decades. Manufacturers, wholesalers, and retailers have tried setting their inventory levels using forecasts based on historical data of past years' sales. But what has changed since the onslaught of the Great Recession is that business enterprises have worked very hard at making their supply chains more efficient in a drive to align inventory with consumer demand. As a result of that early, albeit modest success, there are fewer spikes in production.

Supply chain synchronization does not show up in macroeconomic discussions because it is a relatively new phenomenon. Swings in output are a normal part of the business cycle in classic economic theory. As consumption falls, manufacturers find themselves stuck with large quantities of unsold product. They then curb factory production, laying off workers. They mark down prices to goad consumers into buying products already made. After the excess stock gets depleted, manufacturers then resume production, hiring back workers, to meet customer demand.

That was what occurred in the twentieth century. But in the past few years, the world has changed, calling into question classic economic theory. In the last decade, many big companies have taken advantage of advances in supply chain management practices using special software to manage production and inventory. They have employed inventory optimization software to determine how much stock to keep on hand. They have employed multiechelon inventory software to determine where best to hold that inventory in the supply chain. They have employed network-modeling software to figure where to place the inventory to minimize the amount of buffer stock and still take care of the buyer. And most important of all, they have started to calculate inventory holdings closer to actual demand rather than on the basis of historical forecasts.

As a result, companies have moved closer to matching manufacturing output to true demand, flattening the boom-and-bust cycle of production that took place throughout the twentieth century. With no spikes in manufacturing, a company does not have to hire a drove of workers for extra shifts to run more equipment.

This flattening of the business cycle reduces the need for workers at a time when more and more companies are already relying on software to manage their labor resources and to keep down the employee headcount. Companies use labor management software to measure individual worker activity against preset benchmarks. By boosting workplace productivity, they eliminate the need for additional hiring. Companies use workforce management software to schedule workers for factory, warehouse, and retail store jobs, ensuring that they have just the right amount of workers on hand for the tasks to be done that day.

Along with sophisticated software for improved labor management, businesses continue to pour investment into automation for factories, warehouses, and transportation services to reduce headcount. Increased automation plays a factor in the continuing high rate of joblessness. In fact, a study from the Oxford Martin School in October 2013 said that nearly half of U.S. jobs are in danger of being eradicated in the next two decades due to the computerization and automation. In particular, that study found that jobs in logistics, transportation, as well as in office and administrative support faced a high risk of elimination from automation. In the future, according to that study, the only jobs available to low-skilled workers will be those requiring social skills and creativity that are not susceptible to computerization.

In combination with automation, the improved synchronization of the supply chain will wipe out jobs in factories and warehouses. Synchronization is a trend that's here to stay as supply chain chiefs become more adept at supply and demand alignment. If companies can reduce the lag time in bringing a product from the factory to the market, that means that factories don't have to hire a large number of workers to produce stock that's simply a cushion for a spike in demand. Companies can be run as lean and mean enterprises.

What Western nations in the world are witnessing is structural unemployment caused by a third industrial revolution that enables supply chain synchronization. It's a classic case of what economist

Joseph Schumpeter called "creative destruction." A large group of workers lack the skills and knowledge to secure gainful employment; they're stuck in low-paying jobs at fast-food restaurants and in retail stores. Although it's tempting to take a bleak view of the future, with no jobs for the masses in a mechanized society, it should be noted that the United States and other industrial nations went through this kind of economic transformation more than a century ago. The current economic downturn resembles the Long Depression of the 1880s, when the U.S. economy switched from farming to manufacturing. Back then, farmhands who lost their jobs due to the mechanization of agriculture had trouble finding new work. Likewise, during this current transition period, many low-skilled workers will be hard-pressed to find jobs. As many educators realize, the third industrial revolution requires workers with advanced knowledge in the areas of computer science and engineering.

Surprisingly, few if any economists see the role of supply chain synchronization in the economic transformation taking place in front of their eyes. In my view, that's probably attributable to the fact that they don't look for it. But the evidence is out there. In fact, research on the U.S. economy sponsored by the Council of Supply Chain Management Professionals (CSCMP) provides some evidence that supply chains are becoming more ruthlessly efficient. CSCMP sponsors research that attempts to quantify the impact of logistics on the U.S. economy. The research was started back in the 1980s by the late Robert Delaney as a way to demonstrate the positive impact of transportation deregulation in lowering logistics costs in the United States. Today, economist Rosalyn Wilson continues Delaney's research and updates the study each year.

The *2013 Annual State of Logistics Report* offered some data that shows that some supply chain chiefs are mastering supply chain synchronization. Despite a rise in business inventories in 2012 in the United States, the inventory-to-sales ratio stayed stable in the U.S. economy. That ratio is a measure of a company's on-hand inventory relative to its net sales.

A look back at the inventory-to-sales ratio data shows that in the 1992, that proportion stood at 1.56. Over the course of the next decades, it fell drip by drip, nudging down a fraction of percentage until it reached 1.25 in 2006, just before the nation's economic upheaval. During the nadir of the Great Recession, it shot back up, reaching 1.49 in 2009. Since then the ratio has fallen back, hovering between 1.25 and 1.29.

What that means is that American businesses have become more skillful at determining the right amount of inventory. In particular, retailers have become particularly astute about using inventory management to curtail stock levels, which frees up working capital or cash for the corporate balance sheet. The liberation of working capital boosts business profits. And stocks soar.

What's making this possible, as alluded to earlier, is the wider adoption of the discipline of supply chain management in combination with sophisticated software. The term "supply chain management" first surfaced in the 1980s. The person often credited with the first use of that term is Keith Oliver, who used the phrase in a 1982 article with the *Financial Times*. At that time, Oliver was a consultant with the firm Booz Allen Hamilton Inc. He believed that companies should manage inventory as a single chain of supply across all parts of the entire organization. A Booz Allen Hamilton white paper published in 2003 described supply chain management as "embedded, cross-functional capability designed to unify and rationalize otherwise incongruent parts of a dispersed organization."

Supply chain management was an idea with power. Companies carried too much inventory across all their departments, plants, and warehouses. To reduce inventory, they needed to take a broader view of operations to become more efficient with inventory in serving customers. The problem was that organizations were divided into silos on the basis of functions: procurement, manufacturing, distribution, marketing, and customer service. Silos hinder efficiency. Therefore silos between manufacturing, logistics and distribution had to be torn down so the company could manage the flow from sourcing through production through distribution.

Supply chain management when first proposed represented a new way of business thinking that forced companies to pay attention to the interconnectedness of their operations. It took awhile for chief executives and chief financial officers to understand even modest amounts of inventory create a drag on stock performance. Companies could influence stock prices simply by improving operational performance.

Well-run supply chains also keep their delivery commitments and provide high levels of service, and that drives up repeat customer business and revenue. Companies that embraced and applied supply chain management principles obtained a competitive advantage in the marketplace. Companies that got it got an edge. Still, in the decade following Oliver's first utterance of the term supply chain management, most companies viewed supply chains as simply a cost of doing business. "It was pretty clear that even though [supply chain management] was having a big impact on an organization's service, on its costs, and on the whole idea of response to the customer, it was viewed as a cost center," said Ann Drake, chairman and chief executive officer of DSC Logistics in Des Plaines, Illinois, on the *Voices of the Pioneers* video series chronicling the history of the supply chain management discipline. "I think the evolution from cost center to comparative advantage probably started in the late '80s, but it took well into the end of the '90s before anyone really believed it. . . ."

As more companies understood the value of supply chain management, they began creating new executive positions to focus on supply chain synchronicity. The positions usually had supply chain in the title. On the organizational ladder supply chain, chiefs stood a rung above executives in manufacturing, procurement, and logistics. By 2005, the practice of supply chain management had become so mainstream that the professional organization Council of Logistics Management Professionals (CLM) (founded originally as the National Council of Physical Distribution Management) changed its name to reflect the recognition of the concept in business circles. CLM became the CSCMP.

When supply chain management was first proposed, its focus was on streamlining the internal flow within an organization. But in a global economy, supply chains are really extended enterprises. They involve multiple partners working together to bring a product from start to finish to market. The scope of supply chain management expanded to encompass a group of companies.

As academicians began publishing studies on the critical role of supply chains in creating business efficiency and business management consultants developed proven methods to implement strategies, evidence started to appear that showed synchronization of supply chains could improve shareholder value. Well-run supply chains save costs and boost revenues. Improvements to the bottom line and the growth on the topline maximize business profits. A 1995 study of 225 companies done by the consulting firm PRTM in Weston, Massachusetts, found that manufacturers with best-in-class supply chains had cost reductions of up to 7% of revenue in comparison to the average operation. (PRTM has since been acquired by PwC.)

Even though chief executives and boardrooms began embracing the idea of supply chain management in the 1990s, it took a decade or more to completely turn that concept into full-fledged action. Because corporations are generally large enterprises with thousands of employees, it takes considerable time to change organizational behavior. It takes time to introduce a new tactic or strategy and get it to work the right way. By the time the U.S. economy turned down in 2007, a number of companies in manufacturing, retailing, and wholesale distribution had reached a high degree of supply chain execution. In fact, as the world economy teetered on collapse, corporate executives did everything they could to apply supply chain management practices and techniques to extract money from operations and stay afloat.

Managing global supply chains is a complex undertaking that would not be possible to do effectively without computers and software. And it took awhile for the software to come into existence to support supply chain management strategies and practices. Although mainframe applications for business management existed prior to the personal computer revolution, including a few for supply chain activities, it was really during the mid to late 1980s that a group of niche software vendors took advantage of the increased computing power of desktop computers to write applications specifically for various aspects and areas of the supply chain. Vendors making software offerings explicitly for the supply chain were dubbed best-of-breed. That label distinguished that class of software makers from the enterprise vendors who offered a broad suite of business applications. Best-of-breed software companies developed applications for supply chain planning, procurement spending, transportation management, warehousing oversight, global trade, and inventory control. Without these kinds of software tools, it would be nearly impossible for any company to implement any sophisticated supply chain strategy. There are simply too many parts and products for a human mind to keep track of.

By the start of the Great Recession, the software tools were there.

It should be noted that companies didn't rush in and buy all these tools at once. Back then, before the advent of cloud computing, companies had to buy a license for the software and install the application on their corporate servers. Plus, the initial supply chain software packages were not cheap, either. The price on each package ran into the thousands of dollars. Deployments of software were also enormous undertakings. A software implementation project could take months as the application often had to be integrated with existing software that the company already had. The integration was necessary for the exchange of information as often supply chain software took data out from other systems. So, on top of the cost for the software license, a company generally had to pay thousands of more dollars on systems integration, hiring experts to set up the software. Once the application was installed, employees had to be trained on how to use it. Installing supply chain software was often a big decision for a company because of the resources involved, such that projects were done selectively over time.

Because so many companies since 2000 have deployed software to implement supply chain management techniques and practices, it's having an impact in creating the "New Normal" economy. In particular, the use of robust software for demand planning and for inventory management plays a huge role in boosting company profits. In large part due to the development of sophisticated algorithms, those two types of software create the tools to better connect supply with customer takeaway. The mathematics behind these types software is extremely effective.

The software can use actual information on customer purchases to make inventory adjustments. In fact, the latest applications in demand planning software now use point-of-sale (POS) data to drive both replenishment and production. Every time a cashier scans a bar code on a product at the checkout counter in the store, there's a data notation of the purchase transaction. That means the retailer knows what's sold and where at each store in its chain.

Retailers today can use POS data as the way to determine which products to ship from their warehouses to replace items that are actually being bought. They can adopt a replenishment approach of replacing a product only after it's bought. It's a practice called "buy one, ship one." That's in stark contrast to the way it was done even a decade ago when store replenishment was based on a forecast that was really an assumption as to what consumers were expected to be buying. The assumptions were seldom correct. They were really just guesses.

Now if the retailers share that actual selling information with manufacturers—and some pioneers are doing just that—then the product makers can adjust their factory production to make items— not according to a forecast—but to what's actually being bought in the store. A business model of make-one, sell-one would result in perfect supply chain synchronization.

There is some evidence that does show that synchronization is starting to happen. One piece of evidence comes from a study conducted by Terra Technology, which makes demand-sensing software. Its 2011 study found that nine leading consumer packaged goods (CPG) companies in the United States reduced their amount of forecast errors through the use of a sophisticated demand-sensing approach. Taking part in the study were such well-known companies as Campbell Soup, ConAgra Foods, General Mills, Kimberly-Clark, Kraft Foods, Procter & Gamble, SC Johnson, J.M. Smuckers, and Unilever. The study covered virtually all items in the North American warehouses of these nine multinational companies. By the way, those nine companies together account for more than one-third of the North American market for consumer package goods, excluding alcoholic beverages.

During the combined calendar years of 2009 and 2010, those nine leading U.S. consumer product goods manufacturers reduced their weekly forecast error by an average of 40%. Some of those companies even saw their error rate drop by 49%.

Now on one hand, these companies have a ways to go to reach a perfect unity between supply and demand; on the other hand, they have made remarkable and dramatic strides. They have managed to reduce forecast error, and that means they have come closer to aligning production and inventory with sales. By reducing forecast error on what to make and ship to the store, they have become smarter about what to produce. It is a step change, but a positive one with tremendous ramifications for the economy. After all, a forecast error means that the right product is not on the shelf when the consumer comes into the shop, and the company looses a sale. Or the forecast error results in unsold inventory that ties up capital and eventually results in discounted sale at a lower profit.

There are other data out there as well to support the argument that companies are managing to keep their inventories leaner. A Morgan Stanley & Co. survey of 500 U.S. and Canadian shippers—conducted in the first quarter of 2013—found that almost half of respondents—46%—planned to maintain current inventory levels and not add extra stock. In an article titled "The End of Inventory" in the April 2013 issue of *DC Velocity* magazine, William Greene, an analyst with Morgan Stanley, was quoted as saying about that survey: "Shippers continue to manage inventories very tightly, with no evidence of any big restocking in the future."

As companies began seeing the financial impacts that result from maintaining lean inventories, a new mindset has taken hold in the boardroom. It's a mindset that's grown out of the economic downturn of the Great Recession, and it's a mindset that's here to stay. Corporate executives have come to believe in the importance and value to their shareholders of supply chain management. They religiously believe that they can control costs and raise profits through improved supply chain management. With that mindset, business leaders will to continue to look for ways to adjust their supply chain operations to balance inventory with demand.

As companies look to achieve a perfect fit for alignment between supply and demand, they will create what I call "protean" supply chains. Protean is a reference to Proteus, a sea god in Greek mythology who could change his shape at will. Proteus had mutability. The word protean means the ability to take on different forms and shapes.

Since the 1990s, supply chain gurus have described the need for supply chains to have flexibility, agility, and resilience in written articles, in presentations at conferences, and, of late, on blogs posted on the Internet. Although a protean supply chain has all those attributes of flexibility, agility, and resilience, its most important characteristic is mutability. It can respond and adapt to changes in business conditions and marketplace demands. It can alter its supply chain capabilities—people, resources, and technology—in rapid fashion to connect supply to demand. In short, a protean supply chain represents the next stage of evolution in the field of supply chain management.

Why will this evolution occur? Companies will strive to create protean supply chains as way to stay atop a fast-paced global economy roiled by online commerce and electronic trade. Companies will take great pains to develop mutability. They will strive for protean supply chains whose shape can alter quickly in response to fluctuating world events, market conditions, economic upheavals, and disruptions occurring in international trade. In some cases, the transformations in shape will require companies to change the number and locations of their factories and distribution centers. But in many cases, no physical action will be required. The changes will be "virtual" in that software will allow companies to make modifications to their supply chain operations and practices to have a different capability to respond to shifts in demand.

Protean supply chains are influenced by 10 major dynamics taking place today: demand signal realization, nearshoring, network design, segmentation, omnichannel commerce, control towers, personalized manufacturing, shared supply chains, software analytics, and the environmental movement of sustainability. How those trends support, influence, and necessitate the development of protean supply chains is what this book is about.

BIBLIOGRAPHY

James Cooke. Educating for a global economy. *Supply Chain Quarterly*; Quarter 4, 2011.

James Aaron Cooke. The solid-gold supply chain. *Logistics Management*; April 1997.

Demand sensing greatly improves forecast accuracy. *Supply Chain Quarterly*; Quarter 3, 2011.

Jack Ewing. Jobless rate in Euro Zone stays at record. *New York Times*; October 31, 2013.

Carl Benedikt Frey and Michael A. Osborne. The future of unemployment: How susceptible are jobs to computerization. http://www.oxfordmartin .ox.ac.uk/publications/view/1314. Oxford Martin School at Oxford University. Posted online August 16, 2013. Paper. September 17, 2013.

Peter Herckmann, Dermot Shorten, and Harriet Engel. Supply chain management at 21: The hard road to adulthood. Booz Allen Hamilton white paper; 2003.

Logistics and transportation jobs could be in peril. *Supply Chain Quarterly*. Published online October 24, 2013.

Mark B. Solomon. The end of inventory. *DC Velocity*. Published online March 8, 2013.

Voices of the pioneers. *Supply Chain Quarterly*; Quarter 3, 2013.

Rosalyn Wilson. 23rd Annual State of Logistics Report. The Council of Supply Chain Management Professionals; 2013.

CHAPTER 2

DRIVEN BY REAL DEMAND

Until very recently a manufacturer had to try and predict what it would have to produce based on the past. It would have to develop a sales forecast based on what had been sold last year or last month. That sales forecast, in turn, was the basis for a production plan for the factory. And the factory's output was the basis for what got shipped to the customer.

But historical sales are not the most reliable predictor of the future. Too often the unexpected occurs in the marketplace. A product becomes fashionable for some unexplained reason or an unforeseen event drives up sales. And consumers are just plain fickle.

Until the decade of the 1990s, forecasting was a solo exercise done by companies all on their own and often without consulting supply chain partners. Then it occurred to some supply chain executives that two heads might be better than one in trying to predict the future. In 1995, retail giant Wal-Mart Stores Inc. and the pharmaceutical company Warner-Lambert (now part of Pfizer) did a test of collaborative forecasting for inventory replenishment along with the assistance of the consulting firm Benchmarking Partners and two software

Protean Supply Chains: Ten Dynamics of Supply and Demand Alignment, First Edition.
James A. Cooke.
© 2014 John Wiley & Sons, Inc. Published 2014 by John Wiley & Sons, Inc.

companies, SAP and Manugistics (now part of the JDA Software Group). The first pilot between Wal-Mart and Warner-Lambert involved a joint forecast for keeping the amount of Listerine mouthwash in stores. Initially, the two companies exchanged pieces of paper to compare their sales and production forecasts. Later, the two companies demonstrated in a computer lab that the Internet could be used as conduit for information exchange. By the way, the use of the Internet to trade messages was a huge step forward back then. Although big companies used electronic data exchange to move information between their computer systems, they relied on value-added networks (VANs) for data translation. Small suppliers often did not participate in electronic data interchange (EDI) because of the costs for VANs.

In 1998, the Voluntary Interindustry Commerce Solutions (VICS) Association got involved in promoting this method, which took on the name Collaborative Planning, Forecasting and Replenishment (CPFR). VICS itself had come into existence only 2 years earlier in 1986. Retailers, textile suppliers, and apparel makers formed the organization to develop a common bar code and EDI standards for the retail industry. With the support of VICS, a number of companies such as Procter & Gamble (P&G), Nabisco, Kmart, and Kimberly-Clark began testing sharing information to improve forecasting. In 2007, Richard Sherman, writing in an article in *Supply Chain Quarterly*, said that more than 300 companies had implemented some form of CPFR. (By the way, in 2012, VICS merged with the bar code standards body, GS1 US.)

The goal of CPFR was to reduce inventory and better align supply with demand by comparing the manufacturer's production estimates with those of the retailer's sales projections. But collaborative forecasting did not turn out be as successful as originally hoped, said Sherman in his article "Why Has CPFR Failed to Scale" (*Supply Chain Quarterly*; Quarter 2, 2007). Sherman argued CPFR failed to gain widespread adherence because retailers and manufacturers have paid too much attention to historical data in their collaboration on inventory replenishment and not enough attention to point-of-sale (POS) data. "The problem with building inventory and production requirements based on a point-in-time forecast of demand at multiple points in the process is that the demand forecast is always wrong," Sherman wrote. "Whether they are developed at the point of sale or at other points in the supply chain network, forecasts are always wrong."

Because the forecast was so often wrong, the manufacturer and retailer were forced into a position of having to push the consumer into making purchases. That meant spending lots of money on marketing programs as the manufacturer and retailer would attempt to lure the consumer into making purchases by using incentives and discounts to encourage spending.

But, as stated earlier, consumers are notoriously fickle. They may not want the product being pushed on them. And when consumers rebel, products sit in the warehouse or on store shelves until their price is lowered so much that the consumer says, what the heck, I'll buy it. Since the purchase comes at a lower price, the retailer and manufacturer don't get the profit margin either wants.

The alternative approach is to let the consumers pull the products from the supply chain they want. Put another way, the consumer drives production and distribution. That was the idea behind such initiatives during the 1990s as quick response (QR) in the fashion and retail industry, and efficient consumer response (ECR) in the grocery industry. At that time, QR and ERC were both developed as counterattacks. Wholesale clubs and superstores had begun to take away sales from general merchandise retailers while mass merchandisers, wholesale clubs, and supercenters had entered the grocery business. As a way to combat the threat to their traditional turf, retailers adopted QR in which the manufacturer and retailer would work together to maintain the right stock when the consumer came shopping. Grocers followed suit after a 1993 study by Atlanta-based consulting firm Kurt Salmon Associates (now just Kurt Salmon) estimated that the adoption of that strategy could result in a potential supply chain savings of more $30 billion for the grocery industry.

QR and ECR were a different way of doing business. At that time, those strategies were a switch from the established supply chain flow where manufacturers made huge batches of product according to a historical forecast and then pushed them through the retail channel onto the consumer. A few times each year, the manufacturer would offer a grocer or retailer a low price to purchase a large quantity of product and take it off the manufacturer's hands. That practice was known as "forward buying." The grocer or retailer would then use special offers to goad consumers into buying more products. Unsold, leftover products from the promotion were then stored to maintain an adequate inventory for the grocer or retailer until the next promotional deal came along from the manufacturer.

For a pull system of inventory to work, it needs a signal for demand. It requires the use of some type of information related to actual purchases as the basis for replenishment. That was the idea behind ECR and QR. Since then, that idea has spread beyond retailing and the grocery business to other industries. Today, the concept of a pull inventory system driven by buyers is called a demand-driven supply chain (DDSC).

The demand signal used can take a variety of forms. It can be information on the stocking level in a retailer's store or the amount of product inventory in a distribution center. John Lash, vice president for marketing and strategy at the software vendor Terra Technology, said other types of demand signals could include product orders and shipments within a manufacturer network or warehouse withdrawals of product from the retailer's network. Other less common but sometimes useful demand signals could include weather forecasts or housing starts, Lash added, as they could be harbingers that certain items (snow shovels or windows) will soon need to be made and distributed.

But the ideal signal involves data on actual consumer purchases. Because of the widespread adoption of bar codes, the black-and-white symbols ubiquitous on every package and box found on store shelves, retailers can get a specific item count every time a cashier scans a product at the checkout counter. Scanned purchases are fed into a computer database and that, in turn, can be used as the basis for recurrent replenishment.

Around the turn of the past century, some retailers and grocers began using information gleaned at the cash register or POS data to determine what to ship from their distribution centers to their stores. British multinational grocer and general merchandise retailer Tesco PLC has reportedly been doing just this since 2005 in the United Kingdom. The company uses special demand forecasting software that collects POS information every 15 minutes from its stores. Using that information, the software recalculates inventory levels for store items. Based on those inventory recalibrations, Tesco makes delivery truck runs during the day to maintain sufficient stock at the store. That way, a customer shopping in a Tesco outlet can always find the item he or she wants.

Using POS data to drive shipments from a retailer's or grocer's distribution center to the store is what's known as continuous replenishment. The merchant can pick a specific product in its warehouse and then truck the items to its stores for restocking on a steady basis.

Although continuous replenishment is a good first step, it only involves the seller. To ensure a steady supply means that the maker of the goods has to be part of the replenishment equation. In order for the manufacturer to make adjustments in what to make and to ship, it needs up-to-the minute store-level data by stock-keeping unit (SKU) from the retailer or grocer. Done correctly, that can benefit both retailer and manufacturer. According to the Boston Consulting Group (BCG), in a white paper article "The Demand-Driven Supply Chain," retailers "recognize that partnering with suppliers can reduce stockouts, improve service levels and boost overall sales and customer satisfaction."

A few retailers in the United States started the practice of sharing POS back with goods makers as early as the 1990s, BCG stated in that same white paper. "In the past, retailers were reluctant to share real-time POS data with suppliers," BCG said. "Now many companies (including Wal-Mart) provide that information."

At present, many retailers have begun providing POS data to consumer packaged goods (CPG) manufacturers, typically at no charge, said Justin Honaman, a partner working in consumer goods and retail for database software vendor Teradata. Many grocery stores also provide POS data as well as information tied to their loyalty program, which offer shoppers discounts on purchases or points that can be used toward future purchases. Unlike POS data, grocers generally charge for loyalty card data, said Honaman, because that information is regarded as "most valuable."

Typically, retail and grocery stores transmit that data to the supplier via EDI from their computers systems to that of the manufacturers. In many cases, retailers or grocers will give that data direct to the CPG supplier, although sometimes the information travels through a third party. Honaman said that it's generally understood that the third party is suppose to provide the raw data for free or for a nominal charge to the supplier.

Leading CPG companies are working with retailers and grocers to incorporate POS data into their planning. One well-known company that's using POS information as well as customer inventory levels as a demand signal is consumer products giant P&G, based in Cincinnati, Ohio. P&G makes a variety of personal care products, pet food, and cleaning agents. In 2013, when P&G announced the worldwide rollout of a software system to support a DDSC, the company's Global Process Leader for Demand Planning, Rafal Porzuck, said in a release about the initiative: "Traditional demand planning is simply not sufficient anymore. New products, new markets and increased economic uncertainty have created pervasive volatility, and today's supply chain requires technology to access and analyze as much information as possible to successfully forecast consumer demand."

Del Monte Foods is another company that has shifted to a DDSC, according to a case study from the software vendor One Network Enterprises. Headquartered in San Francisco, California, Del Monte makes and distributes food and pet food products for U.S. consumption. Because Del Monte was experiencing inconsistent fill rates, high inventories, and low forecast accuracy, resulting in lower on-shelf availability for its retail customers, the food company decided to transform its supply chain. It put in place a network-wide platform for transparency that uses demand signals to establish a pull rather than push process for inventory throughout its supply chain from retail customers to its suppliers. One Network Enterprises, which has a case study about Del Monte's efforts in this area on its website, provided the technology platform, which does use POS data. After the system was set up, One Network Enterprises said that Del Monte went from 98 to 99% store in-stock, and experienced an inventory reduction of 27% and 99% case fill rates.

Over in Europe, Danone Dairy has also started creating a DDSC. Danone Dairy collects POS data daily from 195 Carrefour hypermarkets, according to a case study published in the Consumer Goods Forum report *2020 Future Value Chain: Building Strategies for the New Decade*. Based on the POS data, Danone Dairy then refines its forecasts for each store and for each item every day, taking into account promotions and replenishment orders. As a result of that program, Danone Dairy increased its on-shelf availability of products in the French supermarket from 93% to 98%.

One supplier that's doing DDSCs, which I've written about for *Supply Chain Quarterly*, is Kimberly-Clark, based in Irving, Texas. Kimberly-Clark makes such personal care products as Kleenex facial tissues and Huggies diapers. During the Great Recession, the personal products maker wanted to become "demand-driven" in supplying its retail and grocer customers. After doing a pilot program, Kimberly-Clark began using a solution from Terra Technology that can use POS data for updating forecasts for replenishment shipments. At the time when I wrote my article "Kimberly-Clark Connects Its Supply Chain to the Store Shelf" (*Supply Chain Quarterly*; Quarter 1, 2013), only three retailers were actually sharing data direct from the store with the personal care products maker. For other retailer customers, Kimberly-Clark was still using open orders and legacy demand planning to generate shipment forecasts using the Terra Technology software. Still, those three retailers who were working with Kimberly-Clark accounted for more than one-third of its consumer products business in North America.

The use of demand signals allowed Kimberly-Clark to develop a more precise and granular metric for forecast errors. Instead of measuring forecast error by product categories, as it did in the past, Kimberly-Clark's new metric focuses on SKUs and stocking locations. This metric gauges the absolute difference between shipments and forecast, and it's reported as a percentage of shipments. That metric shows how the move toward a DDSC has improved replenishment. Kimberly-Clark has reduced forecast error as much as 35% for a 1-week planning horizon and 20% for a 2-week horizon.

Better forecasting made it possible for the company to carry less safety stock. The company took out 1–3 days of buffer inventory, depending on the SKU. Less safety stock meant that less of the company's capital was tied up as inventory. In 2013, the company told me that the combination of forecast accuracy improvements and safety stock reductions had trimmed finished goods inventory by 19% over an 18-month period.

Kimberly-Clark, P&G, Danone Dairy, and Del Monte are just a few of the consumer product makers working on creating DDSCs. Most CPG organizations have at least some aspect of DDSC in place centered on specific retailers such as Wal-Mart, Target, Kroger, and Safeway, Honaman said. Still, at that moment, the use of POS data for the demand signal is more prevalent in a modern trade country such as the United States and the United Kingdom. In emerging nations, data from the distributor on inventory levels are more likely to be used for the demand signal, said Lash of Terra Technology.

No matter the type of demand-signal input, the data still have to be analyzed. Special software with sophisticated algorithms interpret the demand signals to update calculations for replenishment and production. A steady stream of demand signals means that forecasts for distribution, inventory, and production can be revised at more frequent intervals—daily, even hourly. Periodic updating of forecasts increases the likelihood that the store will have on hand and in stock what the customer wants to buy.

Besides maintaining higher levels of in-stock goods, demand planning software can offer insights into other aspects of store operations to increase sales. For example, the software can be used to support product promotions and improve space planning in the store. In the article "Building a Better Supply Chain" (*Teradata Magazine*, Vol. 5, No. 3), Ronald S. Swift wrote that demand tools "will help build better forecasts and account for seasonality, regional variations, pricing, promotions, distribution and manufacturing constraints within the supply chain."

Intelligence in the software even allows for predictive ordering. For example, very advanced demand planning software can even use a weather forecast as a demand signal in its inventory and replenishment calculations. Indeed Tesco has already begun using weather data to help plan store replenishments. In England, for example, a spell of hot weather results in increased sales of barbecue meat. Tesco can factor weather forecasts into its replenishment plan to keep plenty of barbecue meat in its stores when the temperature rises.

Although the software tools require intelligence to perform these kinds of computations, the applications require understandable data. In exchanging information between computer systems, trading partners must use the same definitions for products and parts. As odd as it sounds, a manufacturer may assign a different number for an SKU than a retailer. If the manufacturer's computer system can't match the retailer's designation for a product, or vice versa, then a demand-driven approach won't work. That's why the need for common data required for DDSCs further promotes adoption of an information technology concept called "master data management."

Master data management ensures a semantic consistency for the information being exchanged. "Supply chain performance is dependent on consistent definitions of customers, products, items, locations and other master data objects," wrote Gartner analyst Andrew White in his article "Master Data! Master Data! My Supply Chain for Master Data!" (*Supply Chain Quarterly*; Quarter 2, 2013). "When data is poorly governed and inconsistent, supply chains because less competitive and more time and money is spent on managing information between systems and trading partners."

Along with a common lexicon, all parties to DDSC must agree upon a common view of the data—"single version of the truth"—as consulting firm KPMG LLP calls in its white paper, "Achieving a Demand Driven Supply Chain." "Everybody has to work off a single version of the truth, one view of demand, one view of supply," said Rob Barrett, a managing director with the KPMG and one of the authors of that paper. "And when you move to a demand model, you have to agree on what's the demand signal. You have to get to a single version of the truth that no one disagrees with. That's 80 percent of the battle."

A company needs that single version of truth as a way to build its future revenue projections. "We have a one-number system, a single best estimate," said P&G Mark Kremblewski in a video presentation about how and why his company developed global standardization for demand planning. "We drive the company against a single demand plan and that single plan dries financials. It drives what we tell Wall Street. It drives our supply chain."

But supply chain partners will also need a single version of demand to fashion a collective response. It's easier to come together on a unified view of demand if all companies involved use the same information platform for trading data. "If the company and all its trading partners are on the same platform, and are working on a single forecast, updated close to real-time, then you can see the difference," said Gene R. Tyndall, an executive vice president of global supply chain services at the consulting firm, Tompkins and Associates Inc. "Safety stocks go down, duplications in work are reduced, planning horizons are more narrow, and lead time variability is reduced and limited."

A shared platform also speeds up the supply chain response. Because information on what's been sold has to travel from retailer to manufacturer to supplier, Tyndall said it can often take 3 weeks to turn that information into action. "Getting them all on the same platform would save a lot of that lead time," Tyndall said. "And that would get them closer to being able to adjust faster [to demand changes]."

The platform must also allow for rapid movement of information as close to real time as possible. Claudio Knizek, a partner in the BCG, noted that today information often flows in batches and it can take a week or more for information to flow upstream in the supply chain to the supplier. Fast data exchange for sharing inventory information becomes the backbone for any DDSC.

The advent of cloud-based software makes this task easier. That's because the data repository holding the single version of the truth can be stored in the "cloud"—information technology jargon for an off-premise application or computer that can be accessed via the Internet. "The technology has evolved such that you can truly operate these cloud business models," said Barrett of KPMG. "You have different parties that need to operate off the same set of data.

What better place to put this than in the white space between organizations."

The data must also be squeaky clean. If the data are dirty, then the supplier can't trust the information it gets from the retailer. Dirty data come from misreads of the bar codes at the cash register. Dirty data can also occur if the retailer or grocer does not take steps to prevent information for 1 day being combined with that for another day. "Dirty, incomplete, inconsistent, and generally poor quality data remains one of the most intractable problems plaguing most efforts to improve supply chain performance," White said.

To ensure the transmission of quality information, the retailer or its agent is required to scrub the data. In addition to cleansing, the data will also have to be refreshed regularly. That requires the retailer or grocer to update the information about what consumers are buying at frequent intervals. "Most companies trying to implement a DDSC will need to collect and share data on inventory levels more frequently and increase the degree of data granularity they analyze," BCG stated in its white paper. "Effective DDSCs typically require information on levels of finished goods and work-in-process inventory at plants. Those systems also require SKU-level detail on items in stores, on warehouse shelves, and in distribution centers."

Data issues notwithstanding, an information platform shared by all partners fixes two key problems preventing supply chains from making a QR. It addresses the time lag that it takes for information to go from the retailer to the manufacturer. And it eliminates debate about what to produce. "If you're all connected, latency goes to zero," said Barrett. "And if you agree on the author for the demand signal, then demand uncertainty gets reduced."

Ultimately, a DDSC could lead to automated updates in the replenishment and production forecast. As Knizek explained, today the ordering process begins when a retailer puts in an order for 100 cases to a CPG company. Someone at the CPG reviews the request and then signs off on the order. But that whole process gets turned upside down in a DDSC. Instead of reacting to an order, the CPG maker can anticipate just how much product a retailer will need. In a DDSC, the CPG can monitor inventory levels at the retailer's stores and then proactively suggest order quantities to the retailer for maintaining stock shelf availability. "You don't have people manually reviewing and accepting purchase orders," said Knizek, "The speed of material flow increases dramatically."

An automatic adjustment to production and inventory in response to demand is just the first step, however. What the future holds is for supply chain partners working in concert to shape demand in real time. "The company shapes consumer demand for the products it has, and it steers consumers away from the items it doesn't have," wrote Swift, a former Teradata executive in the article mentioned earlier. "By creating a demand-driven strategy, the company shapes customer demand in the marketplace and a more streamlined supply chain."

One company that's doing demand shaping is RS Components, a worldwide distributor of electronics and maintenance products based in the United Kingdom. RS Components has started to take steps to manage customer buying online and to influence purchasing behavior. In 2012, the company did a reorganization that brought together product management and supply chain into a single new function called offer. One of the results of this was the assigning of a demand planner and a supply planner to each product category. The demand planner can keep tabs on sales and inventory in the category. Should a particular product not sell as envisioned, then the planners can trigger through colleagues within the category the setting up of online promotions and similar mechanisms.

When a customer goes to its website, RS Components follows the buyer's "journey" through the site from arrival to checkout, noting the pages and products visited. If a customer buys a power tool, the website can suggest batteries as an accessory purchase.

As of 2012, the company has begun to index its products in the market to further understand the price elasticity for a specific item. It potentially allows product pricing to now be dynamically linked to both market availability and competition.

In a pioneering move, RS Components is also working with a third party to develop an artificial intelligence program to gain insight into what drives product sales for new-to-market products. The artificial intelligence engine will sit over its website to gather data. "We want the ability to understand why some products take off and others don't," explained Andrew Lewis, Global Head of Offer Effectiveness for the company. "We want to try to introduce some predictability into what can sometimes appear to be a random event."

Although the main goal of demand shaping is boosting sales, it could have another just as important and beneficial result—supply chain efficiency. Let's say a company wanted to spread out work activity in a warehouse to keep workers from sitting around idle during lag times or to reduce the need for having extra workers on the clock during peak shipping times. The online promotion could offer customers an item discount if orders are bought during a specific time window. Varying order demand becomes a way to manage distribution workflows. Lowering margins on items through demand shaping might result in a net savings on distribution costs. Warehousing and transportation resources could be optimized while still meeting customer expectations.

Demand shaping would give trading partners a way to smooth out production flows. As noted earlier, in the past, manufacturers engaged in batch production, offering discounts on volume buys to retailers in a practice called forward buying. Forward buying minimized plant downtime and kept the factory humming along. But that's a strategy from the old days when product makers pushed inventory onto buyers. Forward buying runs counter to the whole idea of a DDSC, which revolves around consumer pull dictating production. On the other hand, DDSCs require flexible production, yet many manufacturers are not there yet in terms of quick changeovers. Although manufacturers are moving toward agile processes that support short production runs, most still must produce batches of products in accordance with a plan. Demand shaping would assist manufacturers in that regard.

But flexible production is just one obstacle to setting up demand-driven chains. There are others. For starters, implementing a DDSC requires a different mindset toward information sharing. Many grocers and retailers still have a deep reluctance or even an outright opposition to sharing information. For smaller grocers, there's a concern of additional costs associated with data preparation. I remember a supply chain manager at a regional grocery chain telling me that his company did not want to bear the expense of scrubbing scanned data from the checkout counter to make that information available to consumer goods manufacturers.

In the age of the Internet, I also suspect that many retailers fear disintermediation; the manufacturer bypasses the retailer by setting up its own web storefront and selling its product direct to the consumer. Right now, the retailer owns the relationship with the customer. The retailer can gather information on the shopper to influence purchases. The retailer can also use loyalty programs, offering discounts to regular customers in a bid to pull away sales form a competing store. Threatened by the growth of online commerce, many retail store executives are leery of sharing such critical information with manufacturers. That's because the manufacturer could use store transaction data on its products to gain deeper insight into shopping behavior and use that knowledge to drive up sales with changes to stocking, marketing displays, and promotions. The manufacturer could analyze that data for insights to gain a competitive edge in the retailer's own store. That knowledge could allow the manufacturer to drive up its own product sales at the expense of a competing product maker whose products the store carries. That could cause headaches for the retailer, who must manage relationships with the supplier. Plus all those insights about shopping behavior would be useful should the manufacturer opt to cut out the store as a middleman and sell direct to the consumer.

Fears of disintermediation notwithstanding, the emergence of omnichannel commerce forces a retailer to work more closely with its suppliers. If the retailer decides to take part in e-commerce, it will need to pass on up-to-the-minute information on demand in some fashion to maintain a steady supply of products to support online sales (discussed in more detail in Chapter 6). If online merchants and traditional brick-and-mortar retailers intend to use merchandise held in both warehouse and stores as inventory to fill orders across all sales channels, they will need at least near-real-time demand information to coordinate supply. There's no room for error with omnichannel retailing as the buyer will skip to another website if the e-merchant can't offer the item. Not only do merchants have to know the amount and type of products on hand in all its stocking locations to fill web-placed orders but they also have to be able to maintain a steady resupply to ensure an adequate level of inventory to avoid stockouts. As more omnichannel retailers begin to practice same-day fulfillment or engage in same-day delivery to customers for web-placed orders, they will require regular refreshment of inventory. And that becomes even more challenging if product makers have no idea of what will be expected from them.

Although information sharing provides the basis for a DDSC, it won't work if all the supply chain partners don't have the same focus on what's important. To make DDSCs effective will require the partners involved to establish a different set of metrics that measures collaboration on performance. Manufacturers and merchants tend to rate their performance differently. Consumer product makers measure themselves on order fill rates in the distribution center. Retailers, on the other hand, evaluate their performance with such metrics as in-stock, on-shelf, in-store, and order-to-delivery lead times.

Gartner analyst Steven Steutermann has argued that retailers and manufacturers in DDSC should pay attention to the metric "on-shelf availability" (OSA) as a way to judge performance. That metric gauges whether the store has the right products on the shelf when the shopper walks in the door. The manufacturer can use retailer data to determine if it's maintaining proper replenishment. There are financial benefits to using this measurement, Steutermann asserted in an article titled "How to Differentiate Your CP Supply Chain with Metrics." He noted that based on industry pilots, a three-point improvement in OSA yielded a one-point improvement in revenue growth. For companies with organic growth between 3 and 4%, he wrote a one-point improvement in revenue growth is "highly significant."

On-shelf availability is just one metric that's been suggested for management of DDSCs. Another metric, suggested by some CPG companies, is "service as measured by the customer." Since a DDSC is a pull system, the key metric should be one that grades the response to a customer request. "They [CPG companies] decided that their service level metrics were mostly for their benefit, and not synched up with those their customers are using," said Tyndall.

DDSCs may also force manufacturers and retailers to change how they work together, even necessitating new financial agreements. The retailer may well have to provide incentives or offsets for any extra costs borne by the manufacturer. "Supplier contracts must be modified to guarantee that decisions made to improve the performance of the supply chain as a whole don't hurt individual parties," BCG said in its aforementioned white paper. "For instance, if the demand for widgets exceeds the supply because of a successful promotion, the retailer will want the product manufacturer to boost production if it requires extra costs, such as additional labor or overtime. If analysis shows that the total incremental profit of selling more units is greater than the total loss for ramping up production, then producing more products is the right decision."

As manufacturers and retailers work in concert to meet true demand, they will have to change workflows and standard operational practices. Take procurement, for example. Instead of buying raw materials or components at the lowest price, finished goods manufacturers may have to shift their procurement thinking from one of purchasing materials to one of purchasing the production capacity to make products quickly in response to changes in demand.

IDC Manufacturing Insights analyst Bob Parker explained this concept of capability-based sourcing (CBS) in an article aptly titled

"Purchase Capacity, Not Products" (*Supply Chain Quarterly*; Quarter 2, 2009). "Traditional sourcing and procurement is based on a familiar path," wrote Parker. "An engineering parts list is translated to a bill of material. Since many of the items are not standard (that is, they can't simply be bought out of a vendor's catalog), the procurement organization sends out drawings to a group of potential vendors, who then bid on producing the specific SKU. The lowest bid from a qualified vendor usually gets the production job."

But with CBS the manufacturer reserves capacity from the supplier to make something, which will be determined at the point when the demand signal can specify the exact part or component that's required. "The capacity reservation may be in the form of a cash payment, or it may be in the form of a binding guarantee that the supplier can use as collateral when it borrows money. The process may be extended such that the supply chain captain furnishes those vendors with raw materials (such as steel, plastic, or even energy), allowing it to utilize its buying power and better hedge price volatility," wrote Parker, a group vice president at IDC Manufacturing Insights. "The CBS approach is attractive in this capital-constrained environment because suppliers that receive contract guarantees have some level of assured capacity consumption and therefore can better finance their operations."

A DDSC will impact logistics practices just as much as procurement. Although at first thought one would think that DDSCs would lead to more frequent, smaller-sized shipments, and a decrease in large-sized loads, that will not necessarily be the case. If companies are staying on top of demand signals, recalibrating production and distribution based on the current buying, they will actually be better positioned to fine-tune supply chain flows.

As a result, companies would have less need to make emergency shipments hauling products on short notice. Companies will likely wait until a certain threshold is reached before undertaking replenishment. That would allow the manufacturer to still ship truckloads rather than partial ones. In fact, one company with a DDSC was able to take advantage of improved inventory forecasts to create mixed-product shipments to its retail customers as opposed to making dozens of single-product shipments. Not only did those mixed shipments cut order lead times and result in more frequent product replenishment, but they also actually allowed the company to consolidate shipment deliveries, thus generating a savings on freight shipping costs if done right.

But DDSCs don't just apply to consumer goods and retail. In the industrial sector, makers of automobiles, medical devices, and construction equipment have as much a need to base their production on real demand rather than forecasts. They, too, can minimize inventory holdings and tailor production by getting real-time information on what's selling. Barrett said companies such as Cisco, IBM, Motorola Solutions, and Seagate have all established DDSCs. "You can apply the principles of demand-driven supply chains to any manufacturing company," he said.

In a fast-paced global marketplace, matching supply with demand becomes crucial to both the survival and prosperity of all companies comprising an extended supply chain. Every company in the extended supply chain—supplier, manufacturer, distributor, and retailer—has a vested interest in combating margin erosion for finished goods. If the supplier wants to be paid at full price for its product, then the retailer must get full price from the consumer or buyer. They both have margins to protect. To do that, they'll have to make sure that the right product gets made and shipped from the factory to the showroom or the store shelf. The product has to be there when the consumer is ready to buy. If the retailer or grocer has to repeatedly discount or lower prices to move the product off the shelf and into the shopping cart, then the manufacturer will be forced to share in the price cut.

So despite any unease by merchants about data sharing, suppliers, manufacturers, distributors, and retailers are all bound together in a symbiotic relationship. They all have a mutual interest in keeping each other's inventories as lean as can be. Not only do DDSCs promise more accurate forecasts down to the SKUs at each store and distribution center in the supply chain but also this approach lowers costs because trading partners can hold less inventory overall in the channel. And it's not just products. Because suppliers have an end-to-end view of demand, they could hold less raw materials and components. Produce-to-stock manufacturers especially cannot afford to maintain high levels of buffer inventory; capital is simply too expensive and demand shifts too often for supply chain partners not to work together. For extended supply chains, a demand-driven approach simply offers too many benefits, too much money to be had.

Not only do DDSCs lower costs but they can also drive revenue by keeping in stock the exact product that the customer wants. That means less lost sales. Tyndall said he's seen as much as a 10% sales increase in a product category on a store level from companies implementing a DDSC. "The 10 percent sales bumps are real if the out-of-stock performance is lower," he said. "And there is no real reason to be out-of-stock or not on-shelf if the entire supply chain is working in synch on the same near real-time information."

DDSCs also boost cash flow as it enables faster turnover of inventory. In its white paper, the consulting firm KPMG asserted that a demand-driven approach "achieves a balanced cash flow through increased sales, reduced operating expenses and working capital improvements." Not only do improvements to fill rates and lower out-of-stocks for goods result in higher revenue but also the reduction in information latency (from refreshed data) results in companies carrying less overall inventory. Automation of the replenishment process can also reduce operating expenses and allow buyers and planners to manage by exception.

Based on his firm's work with clients, Barrett said that companies with DDSCs on average experience a 1–4% improvement in sales, a 5–10% reduction in operating expenses, and a 20–30% reduction in inventory. He also said that DDSCs help a corporation achieve a goal of freeing up working capital.

At the moment, it's fair to say that only a few pioneering companies have implemented DDSCs and then it's usually not across all product lines or all geographic markets. But more companies are recognizing the importance of adopting this approach. "As companies have downed inventory, leaned out their operations, become more efficient, they find that their supply chains are no longer as responsive as they need to be," said Knizek. "A lot of companies are viewing this as the next big thing."

And well it should be. The rise of e-commerce in the Western world requires a dynamic pull-and-push system for inventory. That's because e-commerce quickens the exchange of money for product, part, or a service. Even though the buying part of the transaction is quick, done with a mere a click of the keys on the computer, the physical exchange of goods still is not. Indeed, if all those digital buys happened so fast, it's possible for web-placed orders for an item to snowball and exceed supply. A surge of e-commerce orders can easily outpace traditional capabilities for both manufacturing and replenishment. Although demand surges in the marketplace have happened throughout history, say, from a weather event spiking up consumer purchases for a product, e-commerce by its very nature creates the potential for far more disruptive supply shortages, far more havoc.

Demand sensing is the only way to deal with the potential problem inherent in e-commerce. That's because sophisticated software allows for a degree of anticipation. Partners in a supply chain could spot the trend in the making and fashion a coordinated response to have the stock on hand to meet online orders. Or partners in the supply chain might also be able to manipulate or shape the demand, where possible, encouraging buyers to take the current product in stock to help steady supply flows and to maintain an equilibrium in factory and warehouse operations.

A supply chain that behaves in this manner to demand fluctuations is protean. In accordance with demand, it regularly recalibrates production, recalculates inventory holdings, and reorders replenishment. To do this, as described earlier in this chapter, all companies making up the extended supply chain—supplier, manufacturer, distributor, and retailer—will have to share data on product sales even more quickly than what's being done now. Along with demand information, the partners will have to have greater visibility at the item level of where products and parts are flowing in the supply chain. Up-to-the minute knowledge of product whereabouts would enable supply chains to redirect flows to meet changing demand.

That's why many supply chain experts are bullish on radio-frequency identification (RFID) tags as a way to provide this visibility. If RFID could ever take hold for item-level marking on a wide enough scale, that technology would also provide an alternative to scanning bar codes as a demand signal. RFID consists of a computer chip and an antenna. The tag transmits the encoded data, such as a product's serial number, product, and shipping information to a reader. The advantage here is that no line of sight is required. To read the information contained in a bar code, a laser scanner has to be aimed directly at the symbol. RFID tags, on the other hand, chirp. They send out a signal to convey information about where the item or box carrying the tag is located. In essence, the tag says, "Here I am, notice me." Hence, the tags can confer visibility on products, cases, or the individuals for a location at a point in time.

When RFID first came into prominence, there was widespread belief that it would create end-to-end visibility for products moving through the supply chain; RFID evangelist and pioneer Kevin Ashton coined the term, "the Internet of Things" to describe a world in which everything would have a tag on it, giving any object its own Internet protocol (IP) address. If every product on a truck, in a warehouse, or in store had an IP address, it could be tracked across the World

Wide Web. The Internet of Things would make it possible for supply chain partners to monitor supply chain flow in real time, keeping up-to-the minute tabs on the whereabouts of parts and products.

Although RFID holds out the promise of supply chain visibility, the use of the technology remains limited. The main reason for that is cost. The Automatic Identification and Data Capture team at business consulting firm Frost & Sullivan, which follows this technology, gave me some prices for an article I wrote in 2013. The team said that so-called passive tags (without battery) cost between $0.40 and $20, while active RFID tags (battery as a power source) go between $10 and $50. Although passive tags do provide wireless transmission of data, active tags are preferred for supply chain visibility because they can send out a steady signal.

Bar codes, on the other hand, are cheap, costing a fraction of a penny for ink and paper. That's why RFID has been used only for the most part on high-margin products like fashion clothing or footwear in retail. In addition, the expense of the tags in those cases are usually justified on grounds of theft deterrence; if a shoplifter walks out the store door with a tagged item under his or her coat, then an alarm gets sounded. Electronic article surveillance rather than supply chain visibility has pushed adoption of this technology in retail.

As the price for RFID units keeps coming down, more retailers are likely to embrace this technology for item-level identification at least on products whose price can cover the tag costs. In addition, as more retailers get involved in omnichannel commerce (a subject discussed in more detail in Chapter 6), they may be forced to deploy more tags to enable the in-store visibility of their inventory to meet commitments to service online customer requests, if they want to fill those orders with stock from the store.

But RFID tags and POS scans are just two types of demand signals that companies will use in the coming decade. As companies develop sophisticated predictive software, web sentiment—expressions of product interest on a Facebook page or a blog posting—could even become a demand signal, especially in the case of a new product introduction. Web sentiment could be used in demand planning calculations as to how much product should be made. Companies could also use weather reports as well for demand signals. Weather reports could indicate that snow shovels or flashlights need to move in mass to stores in an area as an oncoming storm approaches.

The need for companies to meet relentless fluctuations in demand in a global economy is what's driving the rise of protean supply chains. DDSCs will force companies to rethink where they source the product from. DDSCs will force companies to prioritize certain customers. DDSCs will require constant reshaping of their networks. DDSCs will force companies to undertake deeper data analysis to glean insights. DDSC will force companies to engage in personalized production. DDSCs will require control towers to provide visibility from start to finish in the global supply chain. DDSCs may even require resource sharing among supply chain partners.

Without mastering demand, companies will fail in their efforts to compete in omnichannel commerce or to absorb the higher costs from sustainability initiatives. Without mastering demand, companies will be stuck with products that consumers don't want, whether shopping in a physical store or online. In short, if companies don't figure out how to master the DDSC of the twenty-first century, they will fail to make profits and will ultimately go bankrupt.

All those points will be explored in more depth in the following chapters in this book. But the key point is this: A protean supply chain, just like any organism in an environment, must be able to respond to stimuli about what's taking place in the world around it and to take appropriate action. Protean supply chains can quickly reassemble their capabilities—technology, people, and operations— to make the right response.

For that to happen, companies connected together in a supply chain must share data on demand and in real time. In today's global, digitally connected world where demand changes faster than ever before, companies will have little choice but to take their data relationship to the next level of intimacy. Without such intimacy, they will not be able to run protean supply chains.

BIBLIOGRAPHY

The Boston Consulting Group. The demand-driven supply chain. *BCG Perspectives*; 2013.

James A. Cooke. The next big things: Control towers and demand shaping. *Supply Chain Quarterly*. Published online June 20, 2011.

James A. Cooke. From demand signals to demand shaping. *DC Velocity*. Published online July 18, 2011.

James A. Cooke. Kimberly-Clark connects its supply chain to the store shelf. *Supply Chain Quarterly*; Quarter 1, 2013.

James A. Cooke. Demand-driven supply chains demand new metrics. *Supply Chain Quarterly*. Published online September 24, 2013.

Amit Gupta, Kevin O'Laughlin, Rob Barrett, and Frank Kang. Achieving a demand driven supply chain. KPMG white paper; 2012.

Mark Kremblewski and Rafal Porzucek. Video: Procter & Gamble: The road to rapid innovation for supply chain excellence demand driven data. Terra Technology newsroom; 2012.

Bob Parker. Purchase capacity, not products. *Supply Chain Quarterly*; Quarter 2, 2009.

Richard Sherman. Why has CPFR failed to scale? *Supply Chain Quarterly*; Quarter 2, 2007.

Steven Steutermann. How to differentiate your CP supply chain with metrics. *Gartner*; April 23, 2013.

Ronald S. Swift. Building a better supply chain. *Teradata Magazine*; 5(3), 14–17, 2005.

Terra Technology's multi-enterprise demand sensing implemented globally by Procter & Gamble. Terra Technology Press Release; June 27, 2013.

2020 Future value chain: Building strategies for the new decade. Consumer Goods Forum, Capgemini, HP and Microsoft. 2011.

Gene Tyndall. Demand-driven supply chains: getting it right for true value. White paper. Tompkins International; February 2012.

Unprecedented collaboration from shelf to supplier. One Network Enterprises Case Study; 2012.

Andrew White. Master data! Master data! My supply chain for master data! *Supply Chain Quarterly*; Quarter 2, 2013.

CHAPTER 3

REGIONAL THEATERS OF SUPPLY

In late 2012 Apple Inc. announced plans to shift some of its computer manufacturing back to the United States. The iconic computer and electronics maker, born in California, had decided to bring back at least a small portion of its production back home from China.

Apple is hardly alone in reshoring—bringing manufacturing back from a foreign country to the United States. Makers of appliances, electrical equipment, lighting, and phones these days are reassessing the value of offshored manufacturing for the U.S. market and operating extended global supply chains, which often entails the shipment of products made in Asia thousands of miles back to buyers in the United States. Motorola Mobility—part of Google—announced that it was going to move some smartphone production to Texas. In fact, The *Washington Post* (May 1, 2013) reported that such well-known American companies as Ford, Caterpillar, and GE had also announced plans to return manufacturing back to the shores of the United States. Other companies that have recently reshored include Master Lock, NCR, Karen Kane, Morey, and Wham-O, according to the organization Reshoring Institute. In fact, a TD Economics report released in October 2012 said that U.S. manufacturing had revived due to the abatement in offshoring of production.

Protean Supply Chains: Ten Dynamics of Supply and Demand Alignment, First Edition. James A. Cooke.
© 2014 John Wiley & Sons, Inc. Published 2014 by John Wiley & Sons, Inc.

But the United States isn't the only Western country that has started to experience a rise in home manufacturing. The *Financial Times* (November 15, 2013) reported on a study by the U.K. government's Manufacturing Advisory Service that found one in six manufacturers in that country brought back production in the past year. In fact, the survey of more than 500 small- and medium-sized U.K. manufacturers said 15% of respondents reshored production, while only 4% decided to shift to offshored manufacturing.

Related to reshoring is another practice called "nearshoring." In cases where the manufacturer doesn't bring back production home to the United States, it often ends up relocating a factory in a country in the American hemisphere close to the U.S. market. Mexico, in particular, has become a popular destination for nearshored production due to its proximity to the United States, especially for automobile production. Chicago Fed senior economist Thomas Klier has noted that American light vehicle production in Mexico has gone from 6% in 1990 to 19% in 2012. The economist has said that Mexico has become an attractive location for car production because of low labor costs, improvements in its infrastructure, and changes in its trade policies.

Auto parts suppliers have followed carmakers to Mexico. For example, JD Norman Industries, which makes metal components, opened a manufacturing facility in 2008 near its customer base in Monterrey, Mexico. "As more of our automotive customers, both OEM [original equipment manufacturer] and Tier I suppliers, opened plants in Mexico, there was a strategic advantage for JD Norman to be near their facilities," company owner Justin Norman told me. "Our customers have aggressively sought to localize their supply base in every region of the world in which they operate."

High-tech companies are also taking part in the nearshoring movement as well. Evidence for that is found in a 2013 UPS Change in the (Supply) Chain survey, conducted by IDC Manufacturing Insights. That survey found a marked increase in interest for engaging in nearshoring among 337 high-tech supply chain executives contacted for the research. Some 27% of survey respondents said they were embracing "nearshoring." What was particularly interesting in the report was that it was the very large and small companies who were the most keen on this type of supply chain realignment. Companies with annual revenues in excess of $1 billion and companies with revenues between $5 and $250 million cited the most interest.

In many respects, returning production to the United States represents a spinning back of the wheel. During the eighteenth and nineteenth centuries, products manufactured in the United States were made and sold for domestic consumption. After World War II, the United States emerged as the manufacturing center for world trade. That dominance was due in many respects to the fact that United States was the only major industrial country whose production landscape had not been ravaged during the global conflict.

Since the late 1970s, the United States has seen a steady decline in its manufacturing base. In 1950, the Bureau of Labor Statistics said 30% of U.S. workers were employed in manufacturing jobs. In 1980, manufacturing employment fell to 21%, in 1990 16%, in 2000 13%, and in 2010, manufacturing was just under 10%. Interestingly, while manufacturing employment has declined from 1950 to now, industrial input increased seven times, as chief economist for the infrastructure advisory firm Moffatt & Nichol, Walter Kemmsies, has noted.

Still, in 1982, when economists Barry Bluestone and Bennett Harrison published *The Deindustrialization of America:* Plant Closings, Community Abandonment, and the Dismantling of Basic Industry, the book's message was considered shocking. America was losing its place as the world's leading manufacturing country. Bluestone and Harrison argued that the decline of American manufacturing would cost America good-paying blue-collar jobs as companies turned to making products overseas.

The trend toward offshored manufacturing observed by Bluestone and Harrison picked up pace in the 1990s. During that period, companies learned that they could reduce bottom-line expenses by shifting production abroad to low-cost countries, the most notable of which was China. In many cases, makers of household-name products subcontracted their production to contract manufacturers. In doing so, they got rid of expensive assets like factories from their books. They could hire contract manufacturers who would have lower, nonunion labor costs. With support from its Communist government, China became the place to go for American companies looking to build products overseas. "The hyper consumer boom of the 1980s, 1990s and 2000 in America fueled the rise of China," said Michael Zakkour, a principal with Tomkins International Consulting in Raleigh, North Carolina, during a December 2013 webcast on reshoring hosted by investment bank Stifel Nicolaus. "You could make products cheaper and still have a huge markup."

Overseas manufacturing is predicated on some critical assumptions. One of these is low transportation cost; a product made in Asia for U.S. consumption must be put on an ocean cargo vessel or flown in an airplane to reach the buyer. In 1986, 2 years after the publication of *The Deindustrialization of America: Plant Closings, Community Abandonment, and the Dismantling of Basic Industry*, the price of barrel of crude oil was below $10. During the decade of the 1990s (except during the Gulf War), crude oil ranged between $25 and $50 a barrel. As world demand for oil increased due in part to burgeoning economic expansion in China and India, prices began rising for oil during the first decade of the twenty-first century after a relative period of stability. Oil went from $50 to $60 to $80 to more than $100 a barrel. On July 3, 2008, crude oil reached a record $145 a barrel. There was even talk that oil-producing countries could not maintain a sufficient output to meet growing world demand. Some were starting to wonder whether Shell Oil geologist M. King Hubbert might have been right when he predicted in 1956 that the world oil production would peak in the year 2000 and then decline, forcing a run-up in prices for those fortunate enough to be able to pay huge sums for that precious commodity.

Rising oil prices ripple through a supply chain. If a steamship line pays more for bunker fuel, then the carrier raises its rate to move maritime cargo unless stymied by intense market competition. If the trucker pays more for diesel fuel, then it requires the shipper to pick up the added cost through a fuel surcharge. In 2008, in a presentation at the Annual CSCMP Conference, Massachusetts Institute of Technology (MIT) professor David Simchi-Levi pointed out that in the United States, every $10 per barrel hike in crude oil prices raises transportation rates four cents per mile. In Europe, a $10 per barrel hike pushes up rates seven to nine cents a mile. I remember him saying that if oil reached $150 a barrel, that would be a "tipping point" that would force companies to start bringing production home from overseas. The high cost of transportation would wipe out any advantage from low-cost labor.

Since the middle of 2011, oil and concomitantly diesel fuel prices have been relatively stable with brief run-ups whenever political events in the Middle East rattle the market. Part of the reason for the relative stability is increased production of shale oil in the United States. There's even talk that the United States could reclaim its spot as the top oil exporter in the world from Saudi Arabia.

Although crude oil prices have fallen back down from the heights of 2008 to just under $100 a barrel, the price for petroleum and thus fuel remains volatile and unpredictable. Events like a hurricane, a

Middle East conflict, or a military coup could send oil prices sky-high again. Since long-distance supply chains are predicated on cheap transportation, companies remain wary about the future price direction of oil and fuel.

That concern about oil prices is reflected in the correlation between imports and oil prices, as was pointed out in a 2014 article in *Supply Chain Quarterly* on shifts in supply chain strategy driven by transportation trends. Authors John J. Coyle, Dawn Russell, Kusumal Ruamsook, and Eveyln A. Thomchik made the following insightful observation: "Year-to-year growth rates of imports from long-distance sources in Asia dropped sharply around the time of the oil price peak in 2008," they wrote, "which is noticeable contrast to a sharp increase in import growth from near-shore sources in North America, and Latin America and the Caribbean during the same time period."

But uncertainty about oil prices isn't the only reason why companies are rethinking offshored production. There are other reasons, and one important one is rising labor wages, particularly in China. In their report, "The End of Cheap Chinese Labor," published in fall 2012 in the *Journal of Economic Perspective*, economists Hongbin Li, Lei Li, Binzhen Wu, and Yanyan Xiong noted that "at the beginning of China's economic reforms in 1978, the annual wage of a Chinese urban worker was only $1,004 in U.S. dollars." That was only 3% of the average U.S. worker wage at the time. By the 2010, the average Chinese worker's annual salary was $5,487 translated into U.S. dollars, according to the *China Statistical Yearbook*. China is becoming a middle-wage country, the authors assert in the article.

In fact, the Boston Consulting Group has forecasted that net Chinese and U.S. manufacturing costs should converge by 2015. Justin Rose, a partner and managing director in the Boston Consulting Group, told me that nominal wages in China have risen 15–20% per year, while similar wages in the United States have gone up very slowly. For China, "productivity has continued to rise rapidly, but less rapidly than the wage increases," he said. On the other hand, in the United States, although productivity has risen, albeit more slowly than in China, it's still high enough to offset any wage gains on the part of American workers.

Another expert who sees an end to the Chinese manufacturing cost advantage is Harry Moser, founder of the Reshoring Institute in Chicago, a nonprofit group set up to promote nearshoring. Moser has said that Chinese labor cost per hour will soon be about one-quarter of the U.S. cost. But Chinese productivity will only be one-third of the U.S. Labor. If raw material cost such as steel and plastic stays the same and total costs get factored in, then Moser has asserted that the total cost for Chinese products sold in the United States will be higher than the total cost for U.S.-made products.

But China isn't the only country that could be impacted by a revival of U.S. manufacturing. Mexico could even loose some of its luster as a destination for manufacturers aiming for the U.S. market. Evidence for that appeared in some reports such as one from Alix-Partners. The global advisory firm AlixPartners has done a low-cost country sourcing study for the past 5 years. In the 2013 edition of the study, AlixPartners reported that America was tied with Mexico as the top destination for companies bringing back overseas production for the U.S. market. Thirty-seven percent of the 137 respondents in the study said they would choose the United States—the same percentage that would choose Mexico.

When it comes to locations for manufacturing, more enterprises are starting to look at the big picture. "A lot of companies made the wrong decision when they offshored," Moser said on the *DC Velocity* webcast "Made in the USA Again: Is *Reshoring Right for You?*" "Repeated surveys suggest that the majority of companies—maybe 60%—of those that offshored looked at only simple, rudimentary cost models," Moser said during the webcast. "They looked at just price or just labor costs and maybe just landed cost and ignored 20 or 30 percent of more of the total cost of offshored production."

The retired president of a machine tool manufacturer, the Charmilles Technologies Corp., Moser has argued that companies should employ a "total cost of ownership" (TCO) analysis to determine the optimal location for a plant to serve a market. The TCO analysis starts with price and then examines 28 other cost factors to compare the total costs of making products in the United States vs. another country. It should be noted that the Reshoring Institute has developed trademarked software—TCO Estimator—that readily does this type of analysis. TCO Estimator as of this writing had built into it the freight rats for shipping from 17 countries. In his article "Time to Come Home?" (*Supply Chain Quarterly*; Quarter 4, 2011), Moser wrote: "When companies focus only on price and labor, they downgrade all other priorities. Companies that employ TCO, however, usually find that almost all of the other costs would favor production close to the end customer."

Costs aren't the only consideration driving more interest in moving production closer to the point of consumption. Nearshoring or reshoring allows companies to fix some other problems as well. When asked what were the merits of nearshoring, in the 2013 UPS Change in the (Supply) Chain survey, supply chain executives from high-tech companies gave some of the following reasons: improving control over quality and intellectual property, diversification of manufacturing due to natural and socioeconomic risks, and skills or technology limitations. In fact, the number one reason, cited by survey respondents, was improving service levels by bringing production closer to the point of demand.

It's not surprising that "increased service levels" was cited as the most compelling reason to nearshore. To date, many companies that engage in "nearshore" manufacturing, making products in countries adjacent to the United States, such as Mexico and Canada, are more interested in the expeditious movement of a product into a store, and thus the consumer's hands. Companies want to reduce "lead time," the period between when a customer places an order and receives the product.

Lead-time reduction was a major reason why Kenai Sports reshored its production of sustainable sportswear. Based in New Britain, Connecticut, Kenai Sports makes athletic clothing from waste and postconsumer content harvested from landfills. When the company first started, it made its apparel overseas in China. But in 2013, the company shifted manufacturing back to Massachusetts, cutting its lead time in about half while improving its ability to monitor quality control. Other reasons for reshoring were to improve sales with quicker sample production and to improve customer satisfaction.

"We recognized that 'made in the USA' wasn't a strong enough value proposition on its own to justify having our customers pay a slight premium for our performance sportswear," Kenai Sports cofounder Charles Bogoian told me. "Operationally, we focused on condensed lead times to allow for greater flexibility in customer purchase decisions. Instead of having to wait the customary 3 months for completely customized apparel, Kenai customers have full assurances that their order would be fulfilled within 45 days. Also, a more responsive supply chain allowed our company to better react to industry changes. Apparel styles and trends are constantly evolving—as a young, growing, company, we always want to have the ability to quickly adapt and take advantage of new opportunities as they arise."

Cutting lead time is the goal behind the concept "speed to market." In fact, a couple of surveys conducted in the past year provide evidence that Kenai Sports is not alone in reshoring for reasons of speed to market. When the consulting firm AlixPartners did a survey on nearshoring in 2011, 25% of the 80 manufacturing company executives taking part in the research said the speed to market was the key advantage to nearshoring. (By the way, 30% cited lower freight costs as the key advantage to nearshoring.)

A 2012 survey conducted by the MIT Forum got similar results. When asked what drove their decision to move manufacturing back to the United States, about 74% of the 198 surveyed manufacturers said it was to reduce time to market for products. By the way, the second most reason given was total landed cost calculations.

Companies concerned about product obsolescence are prone to speed-to-market supply chains. Apparel falls into that category, as clothing must arrive in store in time for seasonal selling, and for that reason, many clothing makers for the U.S. market have set up plants in Honduras and Nicaragua, which, by the way, also offer low labor costs. In Honduras, for instance, apparel manufacturers often set up their factories close to the Port of Puerto Cortes so they can readily truck their products to that harbor for transfer to a steamship line.

Like fashion apparel manufacturers, consumer electronic makers also have products with a short life cycle. Not surprisingly then, a number of electronic makers have set up shop in Mexico to make products there so they can get them quickly into stores in the United States. In many cases, they still source components from Asia. They then bring those parts in through the Port of Lázaro Cárdenas, which is on the Pacific side of the Mexico. Lázaro Cárdenas has a harbor that can accommodate large container ships. The electronic products are assembled in Mexico and are then trucked across the border.

Building a product closer to the target market allows the manufacturer to respond to fluctuations in product demand. The manufacturer can keep less inventory on hand or in the pipeline because no long ocean voyages are required for resupplying products into stores. It's a 6000-mile ocean journey from China to the West Coast of the United States compared with under a 1000 miles for one from Honduras to Miami. Products moved on a ship from Central America take days rather than weeks from China. And products made in the southernmost tip of Mexico can be moved generally in the day by rail or truck to reach the U.S. border.

For companies with products having a short shelf life, it's inventory velocity that matters. The faster the product can be made, shipped, and sold, the more likely the product can be sold at full price and not discounted to move the item off the shelf. One of the keys to success for clothing retailer Zara is its ability to make its trendy apparel on a short lead time and then to sell those fashions at full price in the store, according to an article in *Businessweek* (November 14, 2013).

The concept of speed to market has started to gain appeal and to be adopted beyond companies with short shelf-life products. One reason for that is more companies want to quickly convert their inventory into cash in the New Normal economy. They don't want inventory to sit unsold in a warehouse. Inventory ties up cash. High inventory turnover allows a business to recoup cash more quickly, and for businesses during the Great Recession, the cash conversion cycle became an obsession. The cash conversion cycle covers the span of time from when a company purchases inventory, sells the inventory, and receives a hard-cash payment for the inventory. Generally speaking, a well-run supply chain can speed up the cash conversion cycle.

Shortening the cash conversion cycle in a volatile marketplace requires quick response to demand. Hence, demand-driven supply chains also reinforce a speed-to-market approach for goods production. So, although costs continue to be factors in the location for a plant, market proximity becomes an important consideration in site selection for manufacturing facilities. That's because market proximity could boost revenue and cash flow. Although the reshoring or nearshoring decision generally has revolved around costs, the evaluation needs to start including revenue and cash-flow projections from adoption of speed to market. In short, companies must weigh costs against the potential of speed-to-market revenue in determining global placement of manufacturing locations.

Viewed through the above-mentioned lens, it might no longer make sense to mass-produce garden hoses in China for the seasonal U.S. market (not too many folks buy garden hoses in winter). If the goal is to make the product and then sell all of it in the spring and summer, it could make economic sense for production to take place in Mexico for the U.S. market. Using that same logic, it does not make sense to make garden houses in Mexico or China for France as well. Production for Western European markets for low-value, commodity-type items like garden hoses make more sense to occur in factories in Eastern Europe, which is close enough to consumers and has reasonable transportation distances for shipping. Likewise, if garden hoses are intended for Chinese consumers, the product should be made in China itself or, if wages continue to rise in China, then the manufacturing should occur in a neighboring low-cost country like Vietnam, Cambodia, or Malaysia.

What this all means is that multinational companies making products for distinct markets around the globe will embrace a concept I call "regional theaters of supply." To match product life cycles to consumer demand, to minimize inventory, and to increase revenue and cash flow, companies need to make products as close to the point of consumption wherever the costs are feasible. The near future will see the development of three major supply chain theaters: one for Europe, one for Asia, and one for the Americas. Africa, Latin America, and other regions could all have their own vibrant regional supply theater at some point during this century. "The calculus is where are my customers, what are my costs for logistics in moving product, and what are my costs for production," said Zakkour of Tompkins Consulting on the Stifel Nicolaus webcast about the future direction of reshoring and nearshoring.

Some companies are already moving in that direction. One company that has already done this to some degree is Starbucks. In my article "From Bean to Cup: How Starbucks Transformed Its Supply Chain" (*Supply Chain Quarterly*; Quarter 4, 2010), I described the changes that the coffee company made to its supply chain following the Great Recession. One step the coffeemaker took involved manufacturing products in the region where its product was sold to consumers. In the United States, Starbucks operates coffee plants in Washington, Nevada, Pennsylvania, and South Carolina. But for Europe, the company operates a coffee plant in The Netherlands. For other areas of the world, such as Asia, Latin America, and Canada, the coffee enterprise relies on comanufacturers. Regionalization of manufacturing allowed Starbucks to reduce transportation cost and lead times to its shops.

Another company that's embraced the concept of regional theaters for supply chains is Procter & Gamble (P&G), the world's largest household goods maker. Take one of its products, razors, as an example. It has one razor plant in the United States, three in Asia, two in Latin America, two in Eastern Europe, and one in Western Europe. "Our business model requires that we manufacture near the consumer," Jeff LeRoy, P&G's media relations manager, told me,

"It's just not cost efficient to manufacture consumer goods in one country and export them all over the world."

P&G as of this writing operates 135 plants in more than 40 countries (32 are in the United States). "Our plants export," LeRoy said, "but mostly regionally. For the most part, P&G brands are made in the region where you buy them."

The 2012 study "Change in the Chain" on the high-tech industry, conducted by IDC Manufacturing and sponsored by UPS, provided some initial evidence for the emergence of regional supply theaters. The study involved 125 supply chain executives at high-tech companies, which included consumer electronics and semiconductors. Some 55% of the surveyed executives said that in the next 3–5 years, they plan to diversify the countries where they source from.

Besides questions about their supply network, the study also asked respondents about emerging markets. That's where the study made an interesting connection. Respondents planned to do more selling into countries like India and Brazil, and at the same time, they planned to do more sourcing from India and Brazil. "A third of the respondents [in the study] said they are balancing geographic sourcing," UPS executive Ken Rankin told me. "It's not just going to be North America and China anymore."

A Tompkins Supply Chain Consortium study in 2013 offers further evidence that companies are moving toward regional theaters. The report *Finished Goods Inventory Management: How Today's Outcomes Measure up to Past Results* noted that more companies were starting to handle their inventory on the basis of geographic territories. The report's findings were based on a survey of 65 top supply chain chiefs. Although most companies still manage inventory either corporate wide or on a division level, the number of companies turning that duty over to a set geography had increased from 17% in a prior year's study to 26%. Companies embracing the concept of regional theaters would need to take a geographic approach to inventory.

As companies move toward setting up regional theaters of supply, production and transportation costs will remain huge factors in determining whether the manufacturing occurs inside the borders of the country with the consumer market or in an adjacent country. For the U.S. market, fashion goods will likely be made in Central America as textile production constitutes labor-intensive manufacturing. On the other hand, capital-intensive industry, as noted in a TD Economics study, will set up factories inside the United States. Examples of capital-intensive products, according to TD Economics, are computers and electronics, machinery, fabricated metals, electrical equipment, and plastics and rubber. "We would not see low-value, low-end jobs coming back to the U.S.," said Zakkour. "The supply chain infrastructure does not make sense for that."

Initially, nonunion states in the South will likely be home to this renaissance in reshored manufacturing as producers take advantage of low labor costs and relaxed work regulations in that region of the country. But the advantage held by nonunion states might be short-lived. Dramatic development taking place in manufacturing processes and technology (discussed in more detail in Chapter 10 on personalized production) calls for specialized expertise and advanced knowledge. As factories become more highly automated and use more robots in place of hourly workers for production, reshored manufacturing could well go to states that have the highest portion of engineering graduates. Automated factories require highly trained robotics engineers who will graduate from such educational institutions as the MIT, Carnegie Mellon, and Rensselaer Polytechnic Institute. Along with highly skilled labor, access to low-cost energy may favor one state or locale over another.

Shifts in global consumer spending in the next decade will also play a role in more businesses creating regional theaters of supply. In an article titled "From Comparative Advantage to Consumption Advantage" (*Supply Chain Quarterly*; Quarter 4, 2012), economists Chris G. Christopher, Jr., and Yinbin Li of the firm IHS Global Insight predicted that consumer spending will decline in the United States and Western Europe, while it will rise in China, India, and South America. For example, the two economists said that in 2001, China only accounted for 3% of global consumer spending. By 2012, China is predicted to account for 12.9%. Contrast that with the United States. In 2001, the United States accounted for 36.4% of global consumer spending. In 2021, that percentage is expected to fall to 21.6%.

Christopher and Yinbin believe that supply chains will align their orbits around these shifts in global consumer spending patterns. They even take the view that emergence of strong consumer markets in Asia will keep some factories in that part of the world from being shut down because of nearshoring. They wrote, "Worldwide changes in consumption are likely to result in some production rebalancing during the next 10 years, but those changes may not be as significant as some might expect. It is anticipated that some multinational corporations will relocate production facilities to the United States or elsewhere in the Western Hemisphere, or move operations to Vietnam from China, but this shift will be relatively minor. In addition, the expected growth in the number of consumers and their spending power in markets like China and Vietnam may help to keep production facilities nearby."

Tompkins consultant Zakkour shares the view that many companies will keep a significant amount of production in China to serve the growing market of Chinese consumers. Moreover, China still offers manufacturers an infrastructure as well as access to raw materials and components, which few other nations possess. If anything, manufacturers faced with the prospect of higher labor costs in coastal areas of China will move their plants into the interior of that country. "For size of market, people will certainly want to continue to keep their supply chain and production in China," said Zakkour on the aforementioned Stifel Nicolaus webcast.

E-commerce will play a huge factor as well in the evolution of regional theaters. As more consumers buy products online and expect quick delivery of their orders, it becomes nigh impossible to meet those service levels with an extended global supply chain and to maintain low levels of inventory. E-commerce will necessitate the building of regional supply chain theaters to support online sales (see Chapter 6 for more discussion of this point). Producers will have to position factories and distribution centers near fulfillment centers established to serve urban centers and metropolitan areas. Because proximity to production cuts lead time, it makes it more likely that the retailer could get prompt resupply if demand kicked up for a product.

The emergence of regional supply theaters poses a huge shift in the orbit for supply chains. The United States stood at the center of the world's supply chain orbit during the latter half of the twentieth century. Back then, as the richest country in the world, the United States had the largest consuming market, and the vast numbers of products made around the earth were brought back there. But that won't be the case anymore in the twenty-first century. There will be at least three major supply chain orbits around the globe. A growing middle class in China makes that country the center of Asian regional supply chain theater, and Western Europe will be the focus of another. India possibly could become another. Others could also emerge. This development should have long-term ramifications for world trade.

Indeed, it could well shift trade surpluses and deficits for a number of countries.

The emergence of a regional supply chain orbiting the United States will cause shifts in that country's supply chain flows as companies nearshore and reshore production. Offshored production that commenced in the 1980s resulted in the rise of a dominant west-to-east flow for containerized shipments in the United States. As U.S. companies began making consumer goods in China and other Asian countries, steamship lines transported Asian-made goods to West Coast U.S. ports where they were offloaded onto trucks or trains—often double-stack trains—for so-called intermodal movement into the hinterland or to the East Coast. As more goods are made at home in the states or in adjacent countries, then that west-to-east supply chain flow should moderate. "Currently, Long Beach is a consolidated point of origin for intermodal with more than 40 percent of Asian imports coming through southern California," A.T. Kearney consultant Jeff Ward noted in his article, "Happy Days . . . for Now" (*Supply Chain Quarterly*; Special Logistics Issue, 2013). "If Asian imports declined, replaced by multiple points of origin across the continent, intermodal traffic may become less dense, and service demands more challenging to meet."

While west-to-east flow declines, the south-to-north supply chain flow should pick up. Since Mexico and Canada are the largest U.S. trading partners, there has always been a strong flow of shipments coming from the north and south into the United States. In particular, as more companies set up production in Mexico, supply chain movements from the south should increase. In fact, U.S. Commerce Department data showed a jump in North American Free Trade Agreement (NAFTA) in 2012 from the previous year. In a 2013 article in *Supply Chain Quarterly*, IHS economists Chris G. Christopher, Jr., and David Deull pointed out that American imports from Mexico have gone from 10% in 2009 to 13% by the second half of 2013. Although most Mexican-made products would likely enter the United States on trucks or railcars, some goods, particularly if they're made in Central America, could arrive on vessels at southern U.S. ports.

Still, the biggest change in supply chain flows in the United States will come about from manufacturers that are reshoring. That will result in the creation of multiple points of origin for shipments. Because manufacturers are expected to set their plants up in non-union Dixie states, at least initially, supply chain flows should radiate northward as well as east and west from those southern locations. If supply chain flow patterns shift in the United States, that change has ramifications for the warehousing and transportation industry. In regard to warehouses overall, there should be a need for less storage space as companies making products closer to the target market would be expected to carry less safety stock. The move toward demand-driven supply chains discussed in the previous chapter should result in fewer facilities and a renewed emphasis on throughput or velocity in the remaining distribution centers in the network.

Reshored manufacturing could well result in more short-haul freight movements if plants source materials and components in a region and transport finished goods to regional buyers. For their part, railroads and truckers may find these flow shifts to their benefit. In the past, when Asian-made consumer goods were moved from west to east, there wasn't always the demand for return shipments back to the West Coast. These flow changes may make it more possible for carriers to find ways to bring back their equipment and drivers to the starting point, although doing that may require some triangulation and zigzagged movements on their part.

Regardless of how these flows ultimately play out in the United States, companies will strive to balance production—supply—with consumer demand in an ongoing effort to build demand-driven supply chains for the reasons discussed in the previous chapter. That will result in different supply chain patterns both in the United States and in other nations. As production locations shift—and keep shifting—and the regions for consumption change across the globe, it requires a balancing act that calls for protean supply chains, ones that can quickly alter their shape to respond to the dynamics of the marketplace. Companies will be forced to examine in greater detail than ever before the composition of their supply chains, a subject examined in more depth in a later chapter on network design.

While many Americans may wish to see reshoring from a patriotic perspective and a belief in a U.S. manufacturing renaissance, that relocation of factories is really just part of an ongoing realignment of supply chains in response to global economic forces and market conditions. Changes in global consumption patterns are occurring due to aging populations in Western countries and emerging middle classes with large amounts of discretionary income in countries like China and India. Multinational companies are merely readjusting their supply chain operations to ensure that they will continue to dominate national markets in each and every country.

BIBLIOGRAPHY

Susan Berfield and Manuel Baigorri. Zara's fast-fashion edge. *Businessweek*; November 14, 2013.

Barry Bluestone and Bennett Harrison. *The Deindustrialization of America: Plant Closings, Community Abandonment, and the Dismantling of Basic Industry*. New York: Basic Books; 1982.

Charlie Bogoian. Should you consider "reshoring" despite the cost? *Venture Beat*; March 8, 2013.

Chris G. Christopher, Jr., and Yinbin Li. From comparative advantage to consumption advantage. *Supply Chain Quarterly*; Quarter 4, 2012.

James A. Cooke. From bean to cup: How Starbucks transformed its supply chain. *Supply Chain Quarterly*; Quarter 4, 2010.

James A. Cooke. Supply chain redesigns: Not just about oil. *Supply Chain Quarterly*. Published online April 26, 2011.

James A. Cooke. So near and yet so far. *DC Velocity*. Published online May 20, 2011.

James A. Cooke. All-star analyst. *Supply Chain Quarterly*; Quarter 4, 2012.

James A. Cooke. The changing geography of supply chains. *Supply Chain Quarterly*; Quarter 4, 2012.

John J. Coyle, Dawn Russell, Kusumal Ruamsook, and Eveyln A. Thomchik. The real impact of high transportation costs. *Supply Chain Quarterly*; Quarter 1, 2014.

Michael Dolega. Offshoring, onshoring, and the rebirth of American manufacturing. *TD Economics*; October 15, 2012.

Simon Ellis. 2012 Change in the chain. White paper. IDC Manufacturing Insights; September 2012.

Simon Ellis. 2013 UPS Change in the (supply) chain: High-tech global supply chains: Shifting gears. White paper. IDC Manufacturing Insights and UPS; November 2013.

Chris Ferrell and Bruce Tompkins. Finished goods inventory management: How today's outcomes measure up to past results. Tompkins Supply Chain Consortium; June 2013.

Foster Finley, Russ Dillon, and Jason King. 2013 AlixPartners nearshoring presentation. Stifel Conference Call; April 5, 2013.

Brian Groom. One in six UK manufacturers reverse offshoring in growing trend. *Financial Times*; November 25, 2013.

Higher oil prices call for flexible supply chains. *Supply Chain Quarterly*; Quarter 4, 2008.

Thomas H. Klier and James M. Rubenstein. The growing importance of Mexico in America's auto production. Chicago Fed letter. The Federal Reserve Bank of Chicago; May 2013.

Hongbin Li, Lei Li, Binzhen Wu, and Yanyan Xiong. The end of cheap Chinese labor. *Journal of Economic Perspectives*; 26(4), 57–74, Fall 2012.

Mitch MacDonald. Oracle of the economy: Interview with Walter Kemmsies. *DC Velocity*. Published online February 25, 2013.

Harry Moser. Time to come home? *Supply Chain Quarterly*; Quarter 4, 2011.

Harry Moser with James Cooke. Made in the U.S.A. again: Is "reshoring" right for you? *DC Velocity* monthly editors briefing webcast; June 2012.

Brad Plumber. Is U.S. manufacturing making a comeback—Or is it just hype? Washingtonpost.com/blogs. Posted online May 1, 2013.

David Simchi-Levi. U.S. re-shoring: A turning point. MIT Forum for Supply Chain Innovation and Supply Chain Digest. 2012.

Harold L. Sirkin, Michael Zinser, Douglas Hohner, and Justin Rose. U.S. manufacturing nears the tipping point: Which industries, why and how much. The Boston Consulting Group; 2012.

Chuck Taylor. The of cheap oil: Are you ready? *Supply Chain Quarterly*; Quarter 2, 2007.

Bill Testa. Reshoring discussion. Midwest Chicago Fed Blogs; posted May 2, 2013.

U.S. catches up to Mexico as preferred North American "nearshore" location, survey finds. *Supply Chain Quarterly*. Published online April 25 2013.

Jeff Ward. Happy days . . . for now. *Supply Chain Quarterly*. Special Logistics Issue; 2013.

CHAPTER 4

DYNAMIC LIVING MODELS

During the Great Recession, many companies hit a wall in their attempt to control transportation cost in the supply chain. None of the traditional cost control measures seemed to work anymore. Supply chain and logistics managers consolidated less-than-truckload shipments into full truckloads and then converted truck shipments into intermodal shipments. They switched from air express to truck express for rapid delivery. On international movements, they gave up air cargo and switched their shipments to ocean carriers. Despite all those efforts at cost control, transportation expenses rose, driven in large part by the price of oil, although tightened carrier capacity in the transportation market was also a factor.

With no way to control the two key factors driving up transportation expenses, was there anything a company could do? Yes. Revamp the network. Think of the supply chain as a grid with each supplier, each factory, each distribution center (DC), and each delivery point such as a customer's DC or store as a node or point on that network. The distances between points in the network determine the length of transportation haul. Since carriers as a general rule charge on the basis of miles traveled to deliver a shipment, optimal placement of

Protean Supply Chains: Ten Dynamics of Supply and Demand Alignment, First Edition.
James A. Cooke.
© 2014 John Wiley & Sons, Inc. Published 2014 by John Wiley & Sons, Inc.

each node can shorten the transits from plant or DC to customer, thus reducing shipment costs.

But there are other factors that impact shipping costs besides the length of distance between network nodes. Take inventory holdings—the amount of products, components, or materials that have to be shipped. When transportation prices were stable during the 1990s, it made a lot of sense to set up one central supply location for servicing customers in a region of the world, but not today. In some instances, dispersing stock across a group of regional warehouses may be more cost effective in that decreased transportation charges offset increased warehousing and inventory carrying expenses. Thus, stocking levels and their locations impact the volume required for transportation and thus supply chain costs.

Another factor in shipping costs is the frequency of delivery, and that ties into service levels. Does the factory need a shipment from a supplier every 2 days or would a week suffice? Does the customer need daily replenishment or would every other day be sufficient? Maintaining a buffer stock of inventory does cut down on shipment frequency, but there are other financial ramifications. For one thing, there are additional charges for warehouse storage and handling. On top of that, unsold inventory acts as a drag on company's working capital.

So, although the ideal makeup of a supply chain network could lower transportation costs, any changes to the existing setup could have unintended consequences. Reducing the frequency of transportation shipments could increase stocking levels. Changing suppliers to get a lower component price could raise the risk of parts shortages. For every change, there are trade-offs to be weighed and evaluated.

It's a complex mathematical problem to assess a number of alternatives and to design the optimal structure for a supply chain network that's moving huge volumes of products or parts across the globe. Fortunately, software used for supply chain design and network assessment has become very sophisticated since the first computer programs were written to tackle these types of problems in the 1970s. Two University of California at Los Angeles (UCLA) professors, Arthur Geoffrion and Glenn Graves, are credited with developing the first computerized supply chain network optimization tool in 1972, according to a Supply Chain Video News timeline of 50 years of supply chain progress.

Software applications can model the impact of hypothetical scenarios and on the basis of those simulations make recommendations on courses of action. "The software makes it possible for managers to model, simulate and fine-tune many different supply chain strategies," wrote Simon Bragg, Julian van Geersdaele, and Richard Stone in the article "Seven Signs Your Supply Chain Needs a Redesign" (*Supply Chain Quarterly*; Quarter 3, 2011).

Network modeling software, in particular, has gained acceptance as the application has matured. "Network design is a proven technology and word continues to spread that firms should do this," said Michael Watson, a partner in the software company Opex Analytics and an adjunct professor at the McCormick School of Engineering and Kellogg School of Management. "Companies realize that network design can solve more than just warehouse location problems."

Equally important, network modeling has been shown to be an exercise with a substantial payback. In the article cited earlier, Bragg, van Geersdaele, and Stone asserted that, based on their experience, a network modeling analysis can "identify savings ranging from 12 to 20 percent of total warehousing and transport costs."

The authors of a noted book on supply chain design—the aforementioned Michael Watson along with Peter Cacioppi, Jay Jayaraman, and Sara Lewis—made a similar assertion about double-digit savings from modeling. In their book *Supply Chain Network Design: Applying Optimization to the Global Supply Chain* (FT Press, 2012), Watson and company wrote: "Companies that have not evaluated their supply chain in several years or those that have a new supply chain through acquisitions can expect to reduce long-term transportation, warehousing, and other supply chain costs from 5% to 15%. Many of these firms also see an improvement in their service level and ability to meet the strategic direction of their company."

Tompkins International consultant Gene Tyndall is another expert who also argued that savings from network modeling can be huge. In the Tompkins International white paper "Employing Available Capital Wisely," author Tyndall asserted "supply chain network design can result in substantially improved networks that save millions in total supply chain costs—often anywhere from 5–15 percent."

Network modeling isn't just about costs; it can show companies ways to achieve more revenue. Network modeling can provide evidence that improved order fill rates, reduced order cycle time, and quicker inventory turns could lead to topline improvements. "Doing a network design on just optimizing costs is obsolete," Tyndall said. "You want to use your network to sell stuff. We tell customers that minimizing network costs is not what you want. You may save costs and lose sales."

These tools can be used to solve complex problems affecting manufacturing, sourcing, and distribution. Unlike software that addresses one of those three functions alone, supply chain design tools examine the linkages and therefore can evaluate the ripple impact a change in one of those areas would have on the supply chain overall. Supply chain design software contains algorithms that can take into account multiple parameters and constraints; a particular customer must get overnight delivery or the factory must be kept in a country for political or economic reasons. (Because of the costs for construction and production setup, it's often easier to relocate a warehouse than to relocate the factory.) Most network modeling software tools can factor in lead times and customer service levels for specified accounts—a particular product buyer must get 2-day delivery in refrigerated trucks, for instance. Because the model can weight service and operational considerations against finances, either costs or increased revenue, these applications can determine the optimal number of warehouses and plants required for the supply chain network and then pinpoint their locations. The model thus can assist a company in striking a balance between the costs and sales revenue in meeting customer commitments.

Supply chain design software evolved from applications that appraised networks. In recent years, the scope of network software analysis has expanded to include inventory. As a result of that enlargement, supply chain design applications overlap into an area that was once the province of inventory optimization software.

Traditional inventory optimization software was developed to assist companies in setting targets for inventory at the stock keeping unit (SKU) level. Generally, these applications work by having the user of the tool select a service level. The optimizer engine in the application uses it mathematics to decide the amount of inventory required to service those commitments. It can also provide direction on locations for inventory placement.

Inventory optimization software first came into being in the late 1990s. "The first mathematics for inventory optimization was done in 1958 for the air force but there was no hardware to run it," said Shaun Snapp, the author of two books on the subject: *Inventory Optimization and Multi-Echelon Planning Software* and *Supply Chain Forecasting Software*. "It took decades before computer power was able to keep up with the software."

In the last decade, more companies with mandates to free up working capital turned to these applications as a way to pare down inventory. Industries from high-tech to consumer packaged goods, from heavy manufacturing to pharmaceuticals have used this type of software to set stock targets. "Inventory optimization tools are not really industry specific so they can be used in most verticals," Gartner research director Tim Payne told me for an article I wrote on this topic in *DC Velocity* magazine. "The main distinction between the different tools is whether they can do just finished goods or can go back into manufacturing and consider 'WIP' [work in process] and 'Rms' [raw material] as well."

A lot of the software today used for inventory optimization is "multiechelon," meaning that the applications look across inventory levels at all stocking locations. When first developed, inventory applications looked at stocking requirements for a single warehouse to maintain service levels. Nowadays, this software considers stock held in multiple inventory locations as if all the items belong to a single common pool. As a result, the tool can calculate inventory needs throughout the entire supply chain, helping determine how much stock to hold at the factory, at the regional DCs, and at the central DC. Because the multiechelon solution takes an end-to-end, big-picture view of inventory, its calculations generally result in less stock in the overall supply chain. Although the software works exceptionally well, many supply chain managers still don't trust the results despite the science behind it. "Somebody will look at the output [from the software] and say, 'I'm not going to do this. We're used to maintaining a certain amount of inventory,'" said Snapp.

Companies often use inventory optimization software to determine the minimal amount of safety stock, which is inventory set aside to address variability in supply and demand. In fact, Snapp has asserted that safety stock is the only factor that accounts for variability in the supply chain. In his view, companies should allow supply planning software to calculate safety stock automatically in response to demand fluctuations. In fact, he said a company can set safety stock dynamically—using demand and supply fluctuations—without using inventory optimization software. Although inventory optimization and multiechelon inventory software will produce a safety stock value, the focus of its use should be on optimizing stock levels across the nodes in the supply chain.

The savings from inventory optimization doesn't just come with the elimination of excess stock, though. Optimization can save a company lots of money because of the huge costs associated with carrying parts and products. In the article "Ten Best Practices You Should Be Doing Now" (*Supply Chain Quarterly*; Quarter 1, 2011), consultant Bob Engel of the firm Resources Global Professionals wrote: "The 'real cost' of holding inventory often is higher than the generally assumed 20 to 25 percent. In fact, recent research reveals that inventory holding costs could represent up to 60 percent of the cost of an item that is held in inventory for 12 months. Those findings included the holding cost of insurance, taxes, obsolescence, and warehousing."

One early user of inventory optimization software was Deere & Company, based in Moline, Illinois, or John Deere, as it's more commonly known. Back in 2001, John Deere decided that it wanted to trim inventory as a way to free up capital. In particular, it wanted to recalculate the inventory holdings for its Worldwide Commercial and Consumer Equipment (C&CE) division, which sold such products as lawn mowers, utility vehicles, and golf course maintenance equipment.

The problem facing John Deere was that only about 30% of the product sat in its own warehouses. Seventy percent of the inventory was situated in the showrooms of 2500 independent dealers. Even though the dealer inventory was technically a "receivable," John Deere was carrying the equipment to the dealer interest free for a period of time. A software analysis confirmed management's beliefs that it could get by with leaner inventory. In particular, the analysis concluded the company could take out $1 billion of inventory if it held the correct amount of inventory at correct locations throughout the year. In my article for the *Supply Chain Quarterly* (Quarter 4, 2007), "Running Inventory Like a Deere," Loren Troyer, then director of order fulfillment at Deere's C&CE division, said his company came to the conclusion that inventory "optimization tools" could help his company correct those issues.

Because of dealer concerns, John Deere phased in the inventory reduction program suggested by the software over a 4-year period. The company implemented a more robust planning process at its five North American factories to lower inventory target levels. The factories also reduced lead time, building products closer to the seasonal demand. That meant John Deere had to work more tightly with its suppliers to furnish parts more quickly. As a result of those changes, the order cycle time went from 10 to 5 days.

If less inventory is kept in the pipeline, then the customers—dealers in this case—have to get faster replenishment or they lose out on sales. To facilitate a more rapid order turnaround, John Deere actually added another layer to its distribution network. Along with the five traditional DCs near its plants, the company added five regional DCs. That change meant dealers were more likely to get a product out of a regional DC than the one at the factory. In this case, shipping from regional DCs actually helped to lower freight costs. That's because often when the factory DCs shipped, the trailers were only 85% full on average. The regional DCs hold about 5 days worth of shipments. Even with the addition of regional DCs, the overall inventory dropped as dealers ended up carrying less equipment and got faster replenishment. John Deere was able to cut in half the overall amount of inventory its channel.

During the Great Recession, many companies discovered—as John Deere did—that inventory optimization applications could maximize utilization of their inventory assets. Although inventory applications remain hugely popular for that very purpose, network design software applications are starting to be used for inventory assessment. Despite the overlapping boundaries, some experts contend that the two solutions still differ in many respects and represent entirely distinct solutions. "ND [network design] is at a level much higher than the SKU-location level that IO [inventory operation] operates at," said Sean P. Williams, an associate professor of operations and technology management at the Boston University School of Management. "IO also handles significant complexity in terms of review periods, batching, nonstationary inventory dynamics that ND models do not address."

Be that as it may, most supply chain design software include inventory in their orbit of analysis. It examines the supply chain configuration and inventory holdings in tandem. Manufacturing capacity can also factor into stocking locations. The flow through the distribution network can determine the amount of safety stock that must be held. Customer service requirement can determine stock levels. Because so many factors are linked together, supply chain design software examines the impact of one relationship on another.

Although both types of software take inventory into account in their calculations, Hemant Bhave, director of Supply Chain Strategy Solution at Barloworld Supply Chain Software, said that inventory optimization and network design have two different purposes. In Bhave's view, inventory optimization tools are designed to evaluate risks in a preoptimized network. IO software set stocking levels based on the risks of demand variation, target service levels, transport lead time, supplier reliability, and replenishment cycle. Those variables are not factored into account in supply network software, which seeks to determine the most efficient route given production and warehousing capacity constraints along with available transportation options. That said, he noted that many companies use both kinds of applications "to drive more value in their supply chain operations."

A supply chain design software exercise begins with a company building a base model of its current state of operations or inventory holdings. The time required for a network design project ranges from 3 to 9 months, Tyndall told me. A number of factors can impact the project's timeline. First, there's data gathering. The scope of the project—whether the design is domestic or international—is another factor. The complexity of the business, such as the number of products, markets, and channels, can also impact the project length, said Tyndall.

Establishing a baseline requires gathering information about historical shipments, inventory holdings and locations, manufacturing output, customers, suppliers, and carriers. That information can usually be gleaned from other software systems a company is using. Enterprise resource planning (ERP), warehouse management systems, transportation management systems (TMSs), and order management systems are just a few of the key applications that might furnish information for analysis. To model inbound shipment flow, data might also be needed from purchasing systems. Necessary data would include location of suppliers and customers, frequency of both outbound and inbound shipments, and even the product weights, types, and quantities typically shipped and received into plants and DCs. "More firms have the data readily available, which makes it easier to do these studies," said Watson.

Tyndall added the design project also needs sales forecasts as far out as 3–5 years. In addition to sales projections, Tyndall said it's critical for companies to use "actual" data. To make the model more accurate, companies need information on actual real estate costs supplied from real estate companies, actual freight costs based on current shipping rates, and actual operating costs for taxes and energy use.

By building a baseline or a picture of current supply chain operation, the software can then evaluate proposed changes to manufacturing, procurement, distribution, or customer service. It can compare various what-if scenarios and quantify impact from each scenario.

For instance, a company can quantify the amount of increases in inbound transportation costs should a European plant switch from an Asian supplier to a U.S. one. A company can quantify the impact on order lead times on customers if it created a leaner network with fewer stockholding locations. Simulation allows a company to envision a different setup for the supply chain and gauge the various impacts of an alternative design. "You do design to get at the supply chain you need tomorrow," said Don Hicks, CEO of Llamasoft, one of the many vendors of network modeling software.

Some of the most advanced supply chain design tools can even take into account changes regarding customer demand and commodity pricing. In the article "Painting a Bigger Picture" (*Supply Chain Quarterly*; Quarter 4, 2009), authors Knud Erik Wichmann and Tim Lawrence of the London-based PA Consulting Group said software advances have lead to "supermodeling." They wrote, "Supermodeling . . . not only examines physical production and distribution costs but also takes into account operations planning aspects such as supply management, manufacturing planning and delivery management. In other words, it assesses the impact of the various cost and value drivers, such as labor, transportation, technology and productivity, on the entire network."

When companies are considering a change in their supply chain, modeling can provide an advance peek into what that change would mean for manufacturing, distribution, and procurement as well as company finances. Modeling gives them a way to preview what could happen and to prepare for implementation. Modeling alternative supply chain structures for implementing strategies lets a company get an idea as to what's involved prior to making any moves. On the basis of the results from modeling, a company can move forward with a plan. That could prove the difference between successful or botched adoption of strategies discussed in this book such as segmentation, demand-driven, omnichannel fulfillment, and regional supply theaters.

Keep in mind that changes to the supply chain can be costly as they often involve an initial outlay of capital before receiving any payback. Since the addition of another warehouse or plant to the network involves a capital expense to the business, the software can put numbers on the potential impact. It can show whether the savings on lower transportation costs or reduced inventory justify an investment in building a new facility.

One supply chain strategy that calls for modeling is implementation of an omnichannel approach in which a retailer serves both online and in-store sales from a common pool of inventory. (Omnichannel strategy is the focus of Chapter 6.) As retailers consider ways to put in place an omnichannel strategy, they must answer questions as to whether the number and locations of both existing DCs and stores are suitable to handle same-day delivery for customers in a selected area. With modeling, they can explore what-if scenarios involving the use of multichannel distribution or dedicated web merchandise warehouse for direct-to-consumer order fulfillment. They can simulate the impact of having a facility dedicated to direct-to-consumer fulfillment on transportation costs versus that of shipping those orders from a store location. They can investigate whether a multichannel DC might have a ripple impact on inventory holdings at other warehouse sites. Through supply chain design modeling, they can attempt to calculate the proper allocation of inventory across an expanded network that would include fulfillment centers, warehouses, and stores. They can also look at which locations in the network should be designated to handle product returns from the viewpoints of both costs and customer service. Bhave of Barloworld Supply Chain Software said one of its clients used network modeling to examine its omnichannel strategy. Not only did the customer realize a savings in supply chain from the exercise but also, Bhave said, the company "more importantly was able to find the optimal mix between number of warehouse locations and balance between regular and expedite freight."

Network modeling also can be used in evaluating the impact of segmentation strategies in which differentiated service levels get assigned to individual customers or distinct customer groups. (Segmentation is discussed in detail in Chapter 5.) For companies embarking on supply chain segmentation, they can test how the application of different service levels for good or poor customers might impact inventory placement and levels throughout the network before embarking on that course. They could model the impact on transportation requirements and shipping cost if they chose to provide faster delivery or slower delivery to certain customers. Since segmentation strategies are often designed to maintain or enhance margins, modeling gives a company a way to determine whether changes to inventory or transportation could well cost the business more money, thus undermining the intent of improving profit margins.

"From a customer perspective they [companies] are able to not only assign priorities in terms of different groups of customers, but they also be able to evaluate sensitivities around changing fulfilment levels," said Bhave of Barloworld. "An interesting trend seen in the board rooms is to look at profit rather than cost and let that drive changes in the supply chain. More importantly segmentation by profit and priority leads to serving the biggest and most profitable customers first—80 percent—and then let the costs and capacities dictate what portion of the trailing 20 percent of customers will get served."

Modeling can be used to assess "rightshoring" or "nearshoring" to validate the veracity of assumptions about relocating production. As discussed in Chapter 3, companies should examine a list of factors in deciding whether to locate manufacturing near the locus of consumption. It's not just a simple trade-off of the costs of labor in unit production against transportation expenses. An assessment should look at total landed costs. It should take into account other important considerations such as national taxes, custom duties, tariffs, and inventory carrying costs. It should measure manufacture capacity and assess whether a shift can maintain the right stock levels.

Modeling can provide that "big picture" number to make an astute assessment about where to site production. For a company that chose to offshore production in the first place to a low-cost labor country to save money on manufacturing, modeling could show whether or not a combination of freight savings from shortened transits and lower inventory carrying costs from reduced safety stock balance outweighs the higher labor costs from reshored manufacturing. Modeling could also be used to evaluate various country settings. For a company planning to establish a regional theater of supply, modeling allows it to map the supply chain footprint for a distinct geographic area.

The software can also be used to evaluate strategic sourcing decisions. For instance, a company conducting a review of procurement costs could model the ramifications of changing a supplier or adding suppliers. Modeling might show that, while changing suppliers could yield a savings on material costs, the flipside of change could increase landed costs.

As more companies adopt a demand-driven focus to their supply chains, they can appraise the impact of more frequent replenishment from their DCs and more frequent production changeovers at their plants. They can evaluate whether their network setup can accommodate faster flows of product to their customers. They can use supply chain modeling to determine cost impact from faster product flows.

Another area for supply chain modeling is determining ways to restructure the network to reduce greenhouse gas emissions. As more companies embrace the concept of sustainability, they are looking at supply chain operations as one area in which they can curb the release of carbon dioxide and other greenhouse gases from the use of fossil fuels in manufacturing and distribution. Modeling allows a company to get an estimate as to whether modifications to the network layout or different types of transportation methods can reduce its carbon footprint, the total amount of greenhouse gases created by the company's products and organization. (Chapter 11 takes a more in-depth look at carbon emissions.)

"Carbon emission is increasingly becoming one of the components used to measure the performance of a supply chain," said Bhave. "With Europe forcing targeted reductions in supply chain emissions companies will need to factor this constraint in redesigning supply chains. The current focus though is more for reporting as opposed to as a constraint. More and more organizations are starting to report their global footprint in their annual reports, but this currently can easily be achieved with excel based tools. Network design factoring in carbon as a constraint will come of age as the targeted reductions come to bear through legislation."

When companies first began using supply chain design software, they applied it when they wanted to tackle a specific business problem. For example, if two companies merged, what should the resulting supply chain network look like? If a company plans to introduce a new product into the market, how might that impact transportation carrier utilization and storage in public warehouses? Where would be the ideal manufacturing location to make this new product from a speed-to-market perspective?

In their article, Lawrence and Wichmann described how one company used a model to answer the question: Would a more centralized production and distribution operation structure lower costs without impacting service? The company with that question had a decentralized structure with delivery and service centers in 16 European countries. It wanted to determine which warehouses it could close and what impact such closures would have on service levels from the remaining sites in its network. The company modeled a number of "what-if" scenarios, gauging impact on transportation costs and stock levels. That exercise showed that the company could cut its 16 DCs down to three and receive a 20% savings in logistics costs.

Johnson & Johnson is another company that used modeling software to examine the setup of its European distribution and manufacturing network between 2002 and 2004, according to a case study from the software vendor Insight Inc. The health-care consumer packaged goods company based in New Brunswick, New Jersey, wanted to know whether it was possible to optimize its network, which comprised 12 DCs in seven countries, while maintaining service levels, which were 1-day service for some customers and 2-day service for others. After an evaluation of the facility and transportation costs against service trade-offs, the company reduced its distribution network from 12 to 5 DCs. While transportation costs went up a bit from $6.6 million to $7.6 million, facility costs dropped from $10.1 million to $3.9 million. Insight said Johnson and Johnson ended up with overall distribution network savings of about $5 million.

Another company that used modeling software to assess its network makeup was Carestream Health. After being spun off a separate company, Carestream Health was faced with the question: Was its network the right size to serve customers? Carestream Health was created in 2007 when Onex Corporation acquired The Kodak Health Group from Eastman Kodak. The medical imaging division had been sharing four U.S. warehouses as well as transportation services with its parent, and the change of ownership meant that Carestream Health could no longer use Eastman Kodak's supply chain.

In my article "A Model of Independence" (*Supply Chain Quarterly*; Quarter 4, 2008), Mark Ewanow, worldwide network design and inbound logistics manager at Carestream Health, said his company had initially selected a Colorado location near a manufacturing plant for its nationwide DC. Although that site worked well for products made at the Colorado plant, Carestream Health found itself sending product made in its New York plant out west and then even having to send some product from Colorado back east. It addressed that issue somewhat fortuitously. Due to a serendipitous circumstance, Kodak sold off a warehouse in New York to a third-party logistics company, which was willing to provide storage and handling for Carestream Health.

The medical imaging maker had been using pool points for outbound delivery, a legacy distribution strategy for saving transportation dollars from its days under the Kodak roof. It would ship full truckloads to a pool point, and the shipments would be broken down into less-than-truckload shipments for customer delivery. With its adoption of a two-DC network, Carestream Health wanted to reassess the use of pooling-point locations and to make sure that it had the right locations for keeping delivery costs low.

To determine the optimal pooling-point locations, Carestream Health embarked on a network modeling exercise. It gathered current and historical shipping to feed into the network design tool provided by Llamasoft. The modeling advised Carestream Health to use six pooling points rather than the nine the Kodak Health Group had used. The modeling exercise also validated the assumptions on the locations for the two DCs. Not only were two DCs well situated for its plants in New York and Colorado but they were also well placed for products sourced from plants in Oregon, Mexico, and China.

The network optimization had a big payoff in transportation savings. The right network locations—the combination of two DCs and six pooling points—helped Carestream Health shave $1 million off its annual $50 million transportation budget. Ewanow told me that the savings would have been even greater if fuel costs hadn't been going up so much as they did in 2007.

L.L. Bean was another company that used modeling software to help answer a question: Could it increase throughput without having to build another new DC? Back in 2007, distribution throughput had become a concern for L.L. Bean as its Internet sales expanded. The consulting company on the project, Fortna, used modeling software as part of the assessment of L.L. Bean's multichannel distribution.

In my article "L.L. Bean's Smarter Stocking Strategy" (*Supply Chain Quarterly*; Quarter 4, 2011), Fortna consultant Philip Quartel was quoted as saying the analysis encompassed transportation, capacity, inventory, distribution operations, and SKUs and looked at the impacts of any proposed changes on L.L. Bean's business of selling products online, in stores, or through catalogs. The distribution network study showed that if L.L. Bean made some changes in logistics practices, it could avoid the cost of putting up another DC. One logistics practice change was expanding the use of continuous replenishment. Whereas in the past L.L. Bean had ordered large quantities to keep an item in stock during a selling season, it switched to smaller size but more frequent deliveries from some of its suppliers. It also cut down on the amount of merchandise preparation by shifting those duties to its suppliers.

Although companies historically have applied network design software to solving specific supply chain problems, more companies are recognizing that, in today's dynamic business environment, this type of modeling should be on a regular basis. Constant fluctuations in demand and supply require it. Every time a company brings on a new supplier or customer, it should reexamine its supply chain makeup and stocking points. Every time raw materials or fuel prices go up significantly, a company should reexamine the node composition of their supply chain. "Everybody knows that they're not safe. You just don't know what it is from," Llamasoft CEO Hicks told me during a video interview ("Harnessing Big Data for Enterprise-Scale Optimization"). "Years of volatility, years of pressure have built up. Design offers you a way out of it."

Not surprisingly, many companies are starting to recognize the importance of periodic network modeling. A 2012 study conducted by the Tompkins Supply Chain Consortium found that the average length of time between network designs had gone from 24 to 18 months. In an article in *DC Velocity* magazine on this study, Toby Brzoznowski, an executive vice president at Llamasoft, was quoted as saying, "The concept of designing a supply chain is something you no longer do once. We're seeing design as being a process as opposed to a project."

Periodic reassessment thus ensures that the network structure remains the best it can be in the face of changing customer and marketplace requirements. In a global economy, if the customer demand for a product is starting to shift, say to another region of the country or the world, a reexamination of the supply chain makeup could help companies evaluate whether they have the correct locations to service customers still at the lowest transportation cost. A reexamination would help ensure that inventory is being held at the best locations for a steady supply chain flow, which takes on critical importance as replenishment becomes more demand-driven.

Bhave told me that the shift toward demand-driven supply chains is one of the primary reasons companies are using network design software more frequently. "Historically network design was used once every couple of years," he said. "However, we are seeing customers wanting to review their network at least on an annual basis and in some cases even more frequently." Other reasons for this growing trend of periodic designs include the volatility of fuel prices, exchange rates, and tax structures. Bhave added companies want to factor "changing shorter and medium term growth forecasts so as to better align the network for cost effectiveness."

Periodic reassessments of supply chain design call for companies to take advantage of "living network models." Because the company has already done the work of putting the data into a model of their supply chain, they have a baseline picture of the supply chain operation. That makes it easier to test new scenarios. Because the models will already have been set up, a company will simply have to refresh the data so the simulation can be based on current business conditions. The simulation will be able to quantify the potential impact of changes, say, a new source of supply or a new product launch. In fact, quite a few large companies are starting to have supply chain analysts on staff just for the purpose of running simulations. "Traditional consultants and technology vendors still think of supply chain design as 'network optimization only,'" Hicks told me. "If the only problem you're thinking about is warehouse location, you only need to do that once every one to three years. If you think of design more holistically, you're doing projects all the time."

Living models will lead to scenario evaluations in real time. In other words, a company would do a what-if evaluation before making a decision as to how to serve a customer. That would be done to ensure a high level of service at reasonable costs. Companies would feed real-time order requests into the network modeling software that, in turn, would select the plant origin for an outbound shipment or select the supplier to optimize inbound shipment of parts to the factory. In 2013, Brett Cayot, a global lead for logistics and distribution in the advisory practice of the consulting firm Pricewaterhouse-Coopers (PwC), told me a number of his firm's clients were starting to do just that!

In my article for *DC Velocity* magazine, "The Next Big Trend: Dynamic Optimization," Cayot said that real-time modeling allows a company to select the lowest cost for shipment delivery, whether it's rail or multistop truckload. In the face of changing business conditions in the transportation marketplace, Cayot was pretty confident that this approach could reduce the total number of shipping miles for a business.

If transportation costs were the sole consideration in the selection of an origin plant or warehouse for a shipment, it's possible that a company could use a TMS to calculate the costs of shipping from one factory versus another factory, or from one supplier to another. But a TMS application does not take into account the ripple effect throughout the supply chain. If one plant ships all of its products to one particular customer, that would mean that another plant would have to be prepared to ship a product to a second customer. That dynamic modeling would take into account inventory holdings, stock allocations, and manufacturing capacity as well as transportation costs in determining the shipping origin.

Dynamic optimization gives a company the flexibility to respond to marketplace volatility. If trucking capacity in the United States tightens in the coming decade, as some experts predict, this approach may offer shippers a way to deal with the problem. Trucking capacity is predicted to shrink as more baby boomers retire from truck driving, and motor carriers have difficulty finding replacement drivers. Truck driving is not considered an attractive occupation, and many in the trucking industry wonder how they will fill the seats of big rigs. If a driver shortage emerges in the decade, as most industry experts believe, then shipping capacity for domestic movements by truck could become a major headache for shippers, especially ones seeking to keep a lid on transportation expenses. The availability of a trucker to take a load from a plant or a DC could well become the determining factor for a shipping point. If a shipment must be made from a designated location, as noted earlier, there will be a ripple effect on the entire supply chain. Dynamic optimization will allow a company to make operational adjustments.

As more companies incorporate demand signals to drive replenishment and production, they will need the capability to make those adjustments on the fly at the last minute. That will further encourage companies to engage in dynamic optimization modeling. As described earlier, if a company decides to service an urgent order taking stock from one DC rather than another facility, there's a ripple effect on inventory holdings across the supply chain. Product allocated in one DC may have to be reassigned from one customer to another. A reassignment of inventory has a cascading effect across the supply chain network, prompting not just one warehouse to shift stock allocation but several warehouses to shift stock allocations. The cascade of shifting allotments across the warehouse network could well require different carriers at different costs to service a mix of customers.

Since protean supply chains require companies to have rapid-response capabilities, dynamic optimization and living network models allow for smarter decisions under time pressure. Dynamic optimization and living models help companies to avoid unintended consequences and to make adjustments to sourcing, production, and distribution that keep down costs and raise revenues.

Finally, up to this point, supply chain network redesigns have been exercises undertaken by single companies. But supply chain partners may have to consider modeling the extended enterprise as a joint exercise. As online commerce forces retailers and manufacturers to work more closely together, they both may have to reexamine their networks with the other in mind. If retailers decide to use supplier inventory to fill online orders, they may have to review whether their separate networks are set up to support a coordinated response that ensures fast turnaround and shipping for customer satisfaction. If extended enterprises seek to base the source-make-move on real-time demand, the networks of manufacturer, distributor, and retailer will have to mesh seamlessly to provide product flow. Future supply chain modeling may have to become a collaborative effort done in real time.

BIBLIOGRAPHY

Simon Bragg, Julian van Geersdaele, and Richard Stone. Seven signs your supply chain needs a redesign. *Supply Chain Quarterly*; Quarter 3, 2011.

James A. Cooke. Running inventory like a Deere. *Supply Chain Quarterly*; Quarter 4, 2007.

James A. Cooke. Frustrated with fuel price hikes? Try software tools. *DC Velocity*. Published online August 1, 2008.

James A. Cooke. Redesign the chain to reduce costs. *DC Velocity*. Published online September 1, 2008.

James A. Cooke. A model of independence. *Supply Chain Quarterly*; Quarter 4, 2008.

James A. Cooke. L.L. Bean's smarter stocking strategy. *Supply Chain Quarterly*; Quarter 4, 2011.

James A. Cooke. A network design is never done. *DC Velocity*. Published online September 18, 2012.

James A. Cooke. Getting the big-picture view of inventory. *DC Velocity*. Published online December 10, 2012.

James A. Cooke. The next big trend: Dynamic optimization. *DC Velocity*. Published online April 10, 2013.

James A. Cooke. Missing the potential of inventory optimization. *DC Velocity*. Published online June 10, 2013.

Bob Engel. Ten best practices you should be doing now. *Supply Chain Quarterly*; Quarter 1, 2011.

Don Hicks. Harnessing big data for enterprise-scale optimization. *DC Velocity* videocast; July 2013.

Johnson & Johnson saves $5 million. Insight Case Study. Accessed 2013. Available at www.insight-mss.com/data/InsightcasestudyJ-final.pdg.

Tim Payne. Hype cycle for supply chain planning, 2013. *Gartner*; November 22, 2013.

Shaun Snapp. Why safety stock is not the focus of inventory optimization. *SCMFocus*; March 2013.

Gene Tyndall. Employing available capital wisely. *Tompkins International*; October 2013.

Michael Watson, Peter Cacioppi, Jay Jayaraman, and Sara Lewis. *Supply Chain Network Design: Applying Optimization to the Global Supply Chain*. Upper Saddle River, NJ: FT Press; 2012.

Knud Erik Wichmann and Tim Lawrence. Painting a bigger picture. *Supply Chain Quarterly*; Quarter 4, 2009.

CHAPTER 5

SPLINTERING THE SUPPLY CHAIN

Simply put, all customers are not the same. Some customers generate more sales than others. Some customers generate more revenue than others. Most important of all, some customers provide more profits than others.

It follows then from that premise that good customers should not subsidize bad customers. Obviously, companies should focus their limited resources on customers who contribute the most money to the bottom line rather than those who chip in the least. It makes no economic sense to have a one-size-fits-all supply chain. Since supply chain operations consume considerable company resources, it's only logical that companies should divide their supply chain into segments, giving priority in production and distribution to the best customers or at least matching capabilities to what the customer values the most.

Supply chain segmentation acknowledges the fact that, in a global economy, supply chains have grown more complex as buyers and consumers place inexorable demands on manufacturers, distributors, and retailers. As a result, companies have to make and hold more types of products at a considerable cost and then get those products

Protean Supply Chains: Ten Dynamics of Supply and Demand Alignment, First Edition.
James A. Cooke.
© 2014 John Wiley & Sons, Inc. Published 2014 by John Wiley & Sons, Inc.

more rapidly into the market to satisfy demand. Up against the double whammy of increased product diversity and increased service demands, companies have to hold the line somewhere on costs or at least make informed decisions in weighing the trade-offs regarding service levels for individual customers.

Although many companies have historically grouped products for sale by region or category type, supply chain segmentation involves a methodical, analytical assignment of the service levels required to tend to customers as part of an overall strategy or business objective. As more companies embrace the practice of demand-driven supply chains (discussed in Chapter 2), it's impractical from a cost perspective to frequently readjust production and distribution to demand without some guidance in the form of product and customer classification.

The term supply chain segmentation shows up in business parlance in the late 1990s. British consultant Sean Culey, who helps companies develop supply chain segmentation strategies, said that it was Wickham Skinner in a 1974 *Harvard Business Review* article on the "focused factory" who first pointed out that a "one-size-fits-all" model for the supply chain was not optimal. Culey said Skinner's article is the first reference he knows about in which "someone is talking about segmenting operations."

Central to supply chain segmentation is the idea of value assessment. It was Michael Porter who raised the importance of value creation for business, in his 1985 book *Competitive Advantage: Creating and Sustaining Superior Performance*. In that book, Porter laid out the concept of value chains, a set of activities that companies provide to bring value to a product or service. Although Porter started this important discussion, "companies were still looking at the company as a value chain rather than breaking down the business into VC's [value chains]," said Culey, who's with the firm Aligned Integration Ltd in Birmingham, England.

It was Marshall L. Fisher who put forward the first widely recognized proposal for a segmented model, according to Culey. In his 1997 *Harvard Business Review* article, "What Is the Right Supply Chain for Your Product," Culey said Fisher "described a 2×2 model that help business leaders to define the right approach for the right product."

By 2000, a number of business management consultants had begun talking about the importance of using supply chain segmentation. Culey said Peter Bolstorff was one consultant who was instrumental in the further development of this method. Bolstorff and Bob Rosenbaum authored the book *Supply Chain Excellence: A Handbook for Dramatic Improvement using the SCOR Model.* (The SCOR model is an accepted framework for supply chain processes.) Although Bolstorff and Rosenbaum's book did not specifically focus on segmentation, Culey said it described a methodology where one of the first steps was "to segment" the business into supply chains and then to determine appropriate supply chain metrics.

Another consultant who was involved early on with the development of segmentation practices was Australian John Gattorna, who began writing about this in 1995. Gattorna has argued that companies should use a "dynamic alignment" framework in which multiple, distinct supply chains are matched to distinct, multiple customer groups based on their buying behaviors. "In the past companies have used a 'one-size-fits-all' approach, which implies that all customers have the same buying behavior," he told me. "We got away with this approximation for two decades, which led people to believe it was the way to operate and design [a supply chain]."

In the past decade or so, supply chain segmentation has emerged as a best practice of leading companies. "Segmentation provides a means by which supply chain managers can tailor service agreements with customers to increase sales while reducing operating costs and both fixed and inventory assets," wrote Kelly Thomas, an executive with software provider JDA, in the article "Supply Chain Segmentation: 10 Steps to Greater Profits" (*Supply Chain Quarterly*; Quarter 1, 2012). "It does this by aligning supply chain policies to the customer value proposition as well as to the value proposition to the supply chain as a whole."

The technology research firm Gartner Inc. defines segmentation as "designing and operating distinctly different end-to-end value chains from customers to suppliers, optimized by a combination of unique customer value, product attribute, manufacturing and supply capabilities and business value considerations." Gartner produces a list each year of companies it contends operate the best supply chains on the planet, and segmentation is one of the hallmarks of companies with superior supply chains. Nineteen of the 25 companies on Gartner's list of the top 25 supply chains in 2013 were either designing or running segmented supply chains. "Most companies start out running a 'one size fits all' type of chain, regardless of individual customer requirements," said Gartner's supply chain research director Stan Aronow. "This often results in over-servicing lower tier customers or allowing too many product and support offerings without understanding how these 'services' impact the bottom line."

Segmentation requires a company to understand which supply chain activities offer value to the customer and which supply activities make a profit for the company. The aim of segmentation is to find the intersection between value and profit. The latter means that a company has to understand its costs at the product and customer levels to figure out profitability. Douglas M. Lambert has been making the case for years that companies need to understand the profitability of their customers and their products. Lambert is a well-known academic in the field of supply chain management who's a professor in the Fisher College of Business at Ohio State University. In his article "Which Customers Are Most Profitable?" (*Supply Chain Quarterly*; Quarter 4, 2008), Lambert stated that "managers in companies that have implemented segment profitability reports have been able to identify products and customers that were either unprofitable or did not meet corporate financial objectives."

Most companies can't readily get a quick picture of individual supply chain activity costs for a single customer, never mind a holistic view of all those costs across functions like procurement, manufacturing, and logistics. That's because traditional accounting systems are set up to report aggregate financial information to stockholders and the government. They don't break down costs to the extent required for a customer profitability analysis.

To assess a customer's profitability, companies often start their supply chain segmentation by applying a financial lens to the business, which is typically a "cost-to-serve" analysis. In fact, in his article in the *Supply Chain Quarterly*, Thomas recommended that approach as the starting point for this exercise. Another way to start the segmentation exercise is to first determine the key expectations for each customer and then to gauge the financial ramifications for matching supply chain capabilities to what matters most to the customer. "If a company decides to make changes in the supply chain, it will then use 'cost to serve' to quantify the impact of the changes," said Matthew Davis, a former Gartner analyst who has written extensively on the topic of segmentation.

A cost-to-serve analysis is one commonly used method for breaking down and assigning supply chain costs. The second most common method for calculating supply chain and logistic costs is activity-based costing, which is more often used in financial analysis. Gartner research vice president Tim Payne has noted that activity-based costing does not take into account the constraints and limitations on operations. Cost to serve, on the other hand, does a better job of marrying operation activities to specific supply chain cost drivers.

The concept of cost to serve was first discussed in an academic paper in 1998. Alan Braithwaite of LCP Consulting Ltd. and Edouard Samakh authored an article outlining the concept for the *International Journal of Logistics Management*. LCP Consulting actually holds the registered mark for cost to serve, although Braithwaite told me that lately he prefers to call that method of analysis "net margin management or value-based management."

A cost-to-serve analysis is designed to identify the expense of serving a customer or group of customers. This exercise involves assigning costs for activities such as order processing, order fulfillment, transportation, and warehousing to individual customers or customer groups. The same exercise can be done for product families or delivery channels. The end result is a profile of the cost characteristics for servicing individual customers or customer groups, or products or product families.

Applying this type of scrutiny can provide proof that some customers cost the company money. When costs for service are subtracted against the product price, companies discover that some of its customers provide little or zero profit. Indeed, when companies start assigning all logistics activity costs—transportation, warehouse handling, order processing—a true picture emerges as to what it actually does cost to serve a customer. That picture generally upends assumptions. Companies often discover that customers thought to be profitable in reality don't make the business money at all. Thomas, in his article, stated that 30–40% of a company's customer and product portfolio is generally unprofitable. In his YouTube video titled "Cost to Serve," Rob O'Byrne, group managing director for the management consulting firm Logistics Bureau Pty Ltd. in Sydney, Australia, made a similar claim. O'Byrne said it's not uncommon to find that 30% of a company's orders are so small that when costs are subtracted, there's no profit margin.

Cost-to-serve analyses were historically done with spreadsheets, and many companies still do it that way, putting numbers in the columns and rows of a grid. In his article Kelly pointed out that "leading companies have started with a simple model that assigns transportation, inventory and order costs to products based on their volume and other ordering dynamics."

Companies can create a graph to help visualize where customers stand on the basis of the cost-to-serve analysis. Designate the vertical axis for shipment volume and the horizontal axis for profit margin. Divide the graph into four squares: one for high volume/high profit; high volume, low/profit; low volume/low profit and low volume/high profit. Then, based on the cost-to-serve analysis, assign a customer to one of those four quadrants. But that's only the first step to understanding the customer profile. That's because all products sold to top customers, those in a high-volume/high-profit quadrant, may not yield a profit. That's why companies often apply a second lens or filter to take a deeper look. In the second visualization, products rather than customers are placed into one of those four quadrants. This mapping often shows which products sold to a profitable customer are actually profitable and which ones are not. Those exercises thus provide a graphic visualization of the profit contribution by customer and product.

Doing a cost-to-serve analysis with spreadsheet calculations and the subsequent graphing can be a time-consuming exercise. Fortunately, there's software available to do the job. Gathering the required data can also be a challenge. Although it's possible to put together logistics costs from invoices, if the company wants an end-to-end view of its supply chain that includes manufacturing, getting the necessary information for the analysis often requires pulling data out of other computer systems. Although enterprise resource planning (ERP) systems can't readily generate detailed reports to do the actual cost-to-serve analysis, O'Byrne told me that most companies get the information by extracting product, transaction, and cost data from an ERP system. Davis noted that a key advantage of using software is that a company can change some of the underlying assumptions—say, about customer service levels—and then run the analysis again quickly. "The benefit of a technology solution is that it's more dynamic," he said.

The cost-to-serve method is just one way to apply a financial lens to customers and products in a segmentation analysis. Another consultant who has done segmentation work for companies, Jonathan L.S. Byrnes, takes a somewhat different tack. Byrnes, who's a senior lecturer at theMassachusetts Institute of Technology (MIT), runs the company Profit Isle and has authored the book *Islands of Profit in a Sea of Red Ink*.

Byrnes' starting point is an examination of the profit and loss for every invoice line a company has. Byrnes said he uses that starting point since every invoice has on it a customer and product name. He uses specially developed software that reviews millions of invoices. He said he goes through every item in the company's "P&L" and figures how to assign costs for warehousing, transportation, information technology, and even the "president's parking space." All the sales numbers on invoice lines have to match the company's revenues and the exercise is completed when the fully costed total of the line's net profits add up to the company's reported net profits. The examination takes a month or two to complete.

In his segmentation work, Byrnes said he has discovered that typically 20–30% of the products in the portfolio and less than 10% of the customer base accounts for "150%" of the company profits. "The differential in profitability does not come from gross margin," he told me. "It comes from the difference between gross margin and net [profit], and that's typically supply chain costs. Unless you have a view of profitability, you're working on averages."

Although many companies approach segmentation from the standpoint of cost, companies that have a strong customer focus tend to start their segmentation exercise by classifying customers, Davis said. "People who start with cost-to-serve do so for one of two reasons," he told me. "The business has a cost-reduction target to meet. Or they start with cost-to-serve because they know they are over servicing or underservicing customers."

In contrast, segmenting by customer begins by determining what the customer values most. Some customers want low costs. Some want speedy delivery. Some want product availability. And some might want a product with unique attributes. In his work on segmentation, Goetz Erhardt, a consultant with the firm Accenture in Germany, said a customer needs analysis can be done either by a direct survey of the customer or by conjoint analysis. The latter is a particular type of quantitative method often used in market research. Erhardt told me he uses special software for the conjoint analysis.

Once a determination has been on made as to what a customer or a group of customers values the most, then Davis said a company should do a cost-to-serve analysis to identify the financial ramifications for providing the required service. In his experience, Davis said that companies typically take on supply chain segmentation with a phased approach. The first phase often focuses on product characteristics by completing an analysis of historical orders or shipments—some measurement of product demand—over a defined period of time such as a year. The portfolio of products is mapped into four quadrants based on the total volume and relative weekly variation. For example, certain products can be high volume but have low variation, while others may be low-volume, high variation. Products are then matched to customers. The second phase of supply chain segmentation is identifying customer value requirements such as lowest cost, fastest speed, highest service level, or service differentiation. In segmenting based on the value, the original volume and variation analysis now serves as the product constraint against the desired result.

Once the supply chain has been segmented by cost or value, then a company can develop differentiated service policies for each segment. Generally, these policies address inventory, fulfillment, order promising, transportation delivery, lead times, sourcing, and even manufacturing. Retailers can take also segmentation into account in regard to their policies regarding category management and vendor managed inventory.

Inventory is one critical area where companies should adjust service levels to a particular segment. Higher or lesser amounts of buffer inventory might be kept for different sets of customers. Positioning of inventory at stocking locations could also vary by customer.

A policy could also be set for inventory allocation as well. It could thus dictate that only certain customers when ordering product receive first call on any item in inventory. In fact, a company could make a different replenishment policy for each customer, based on the service required, the volume, and the profitability. A top customer might get the last item in stock, while a poor customer might be told to wait while factory produces the wanted item.

Lead times could also vary. And companies could factor those lead times into the logistic service requirements. Take the case where a customer wants value-added services such as building special product cases for a store display or special modifications to a generic product for a targeted market. Because there's an expense to the value-added service, lead times would be lengthened to meet the customer's priority—product uniqueness. The product maker company might engage in postponement manufacturing, modifying a generic product to the buyer's specifications at the distribution center rather than in the factory.

The policy on selection of a transportation mode and method could also vary by customer. Large retailers may get more frequent deliveries, while smaller ones do not. In addition, only a select group of the most valuable customers would be granted overnight delivery; others would be served by the less expensive but slower forms of transportation such as second- or third-day truck delivery, or even a slower mode like rail or ocean carrier. Cost trade-offs could be taken into account. A customer requiring special packaging might have to settle for lower-cost transportation as a way to maintain profitability for the business.

Although segmentation often deals with logistic services and inventory, customer segmentation can apply to suppliers as well. A customer that values speedy arrival might be assigned to a near-shored supplier that provides short lead times for product delivery, while another customer valuing low price could be assigned to an offshore supplier that results in longer lead times for shipping. To accommodate a last-minute order from a high-margin customer, a contract manufacturer could even shift production schedules or bump up delivery. A factory could provide high-paying customers with make-to-order production, while less valuable customers are assigned to make-to-stock production.

Because segmentation establishes policies for categorizing customers and, in doing so, gives some priority status, this approach requires someone in authority to oversee this practice and to sign off on decisions. After all, service differentiations could prompt complaints from the sales force and jeopardize business relationships when a customer's request is not honored right away. That's why companies adopting this approach should staff and set up a special team to oversee segmentation. A team of supply chain experts with a keen understanding of the customer base can manage customer expectations and impose the discipline required for this strategy to work. If only a select group of customers is permitted to receive overnight deliveries, then the center can manage the political fallout from imposition of this requirement. In a case where a customer's strategic importance to the overall business necessitates favored treatment and a policy override, the center can make that determination.

What benefits do customers reap from segmentation? In a white paper, the consulting firm McKinsey & Co., based in New York, said manufacturers doing segmentation have seen finished goods inventory go down 20–40%, and retailers have improved on-shelf availability 3–5% while seeing logistics costs go down 15–20%. Although cost control may be the reason for engaging in segmentation, this approach can actually increase sales when executed properly. Another consultant told me his clients on average get a 4% increase in revenues from employing supply chain segmentation.

Done properly, segmentation does not have to punish customers who contribute little toward the bottom line. Instead, this approach allows a company to tailor supply chain capabilities to what matters most to a customer. This strategy forces a company to answer itself the question: What does a customer value enough to pay for the service?

So, if a particular customer views low prices as more important than prompt service, then a manufacturer can adjust its supply chain to manufacture in batches and ship products in truckload volumes. On other hand, for a customer that values rapid replenishment—and who's willing to pay for that higher level of delivery service—then the company can adjust production schedules, carry higher levels of safety stock, and even ship products overnight.

Byrnes said, in his experience, companies use differentiated services developed from segmentation to come up with a "game plan" to boost profitability of customers and products. A salesman can go to each customer with specific proposals, such as ordering in less frequent but greater quantities that can make an account go from unprofitable to profitable. Companies should have a "game plan account by account to move them back into profit," said Byrnes. "We can do this for every product and every account."

In some cases, a company might actually find that it has to increase its initial costs for a segmented supply chain to generate more revenue. Davis said one consumer packaged goods (CPG) maker actually increased its logistics costs after a segmentation analysis found a retailer valued on-shelf availability. Although it had to bear the cost for increased deliveries to maintain in-stock positions, the consumable goods maker obtained increased sales as a result.

How long does it take to do the work to set up a segmented supply chain? Tom Craig, president of the consulting firm LTD Management in Lehigh Valley, Pennsylvania, said in his experience a segmentation exercise takes between 4 and 6 months to complete due to the data collection and analysis involved. Once the exercise is done, many companies will then run a pilot for 6 months to a year to validate the segmented supply chain strategy.

One company that has done an excellent job with supply chain segmentation is Dell Inc., based in Austin, Texas, which has different policies for serving consumers, corporate customers, distributors, and retailers. A number of factors drove Dell in 2007 to undertake segmentation as part of a transformation of its supply chain. Back then, there was a major shift in the computer market as sales shifted from desktop to notebooks. While commoditization was pushing product prices down, globalization of the market was driving sales growth outside the United States. Dell was also entering into retail sales of its products. It discovered that what the online consumer wanted—the ability to configure the personal computer to their tastes—did not matter as much to store buyers.

Dell decided to segment its supply chain around customer preferences—speed of delivery, lowest cost of its products, precision, and value-added services. "We designed a segmented supply chain for what our customers valued," said Donna M. Wharton, vice president of Dell's global operations during a *Supply Chain Quarterly* webcast titled "Segmenting Your Supply Chain for Success." The technology provider developed a cost-to-serve mechanism to charge each business unit for the supply chain model and services it used.

In his article, "When One Size Does Not Fit All" (*MIT Sloan Management Review*; Winter 2013), about his work on the project, MIT professor David Simchi-Levi said Dell realized it had to support different types of supply chains; one for build-to-order, one for build-to-plan, one for build-to-stock, and one for build-to-spec. Build-to-order supply chains were suited for online sales with low-volume configurations, while build-to-stock worked best for selling popular computer model configurations online. Build-to-plan fit the retail environment, while build-to-spec was the supply chain approach geared to corporate sales.

Segmentation resulted in a huge payoff for Dell. Simchi-Levi said that the technology maker saw its product availability improve by 37% and its order-to-delivery time cut 33%. Because Dell was able to link costs to what the customers valued most, Wharton further said that supply chain segmentation allowed Dell in a period from Quarter 4, 2009, to Quarter 4, 2011, to see a drop in supply shortages by 57% and a decrease in overdue orders by 80%. At the same time, Dell was able to shift a huge volume of consumer product orders to ocean shipments for a freight dollar savings.

Dell's experience is typical in that companies see service improvements by aligning capabilities more precisely to customer needs. Gartner's Davis told me another story about a consumer goods manufacturer that adjusted service levels after a segmentation analysis of its small store customers in Asia. The manufacturer was selling facial creams, which were low-volume shipments that had high purchase variability. Because the retail outlets receiving these products had limited storage space, they often delayed placing orders and hence were frequently out of stock. Store patrons, however, were paying a premium for the product and expected 100% on-shelf availability.

To service these small stores, the manufacturer changed its packaging and delivery method. It came up with a newly designed single package that held a mix of products. It also shipped these revamped packages daily by courier. As a result, Davis said despite higher unit costs, sales at the segmented stores grew at five times the rate of nonsegmented stores.

But a segmentation strategy does not have to be used for just aligning supply chain capabilities to service customers. It can be applied to managing suppliers and assigning risk to them. In the article "Rules of Engagement: A Better Way to Interact with Suppliers" (*Supply Chain Quarterly*; Quarter 4, 2012), authors Paul W. Schroder and David M. Powell, both with the consulting firm A.T. Kearney of Chicago, argue that companies should segment suppliers according to their capabilities for meeting the company's business value. In segmenting suppliers, a company needs to evaluate the supplier not just on price and quality but also on innovation and reliability. They also need to take into account a supplier's criticality to the business.

Once suppliers are segmented, Schroder and Powell assert that they should be managed "differently based on the business objectives" they help a company to achieve. This necessitates creation of an "engagement model" that defines interactions with the supplier. In turn, the model leads to a plan specific to each supplier. These plans should spell out how the supplier can help ensure a steady flow of parts or products in the event of a supply chain disruption.

Although supply chain segmentation seems to improve supply chain efficiency, whether the orientation is toward suppliers or customers, it's still considered a hallmark strategy for leading-edge companies that make Gartner's top 25 list of those businesses with the best supply chains. "Many of the companies on our Top 25 Supply Chain List have reached the point in their journeys where they are now able to create and sustain this type of differentiated operations," said Aronow, "while still enjoying the synergies of supply chain as a shared service."

The high-tech and consumer products consumers are the farthest ahead in adopting segmentation for their supply chain structure,

Davis said, although the pharmaceutical industry is quickly catching up. The 2013 Life Sciences Supply Chain Benchmarking Study backed up Davis' assertion. That study of more than 200 pharmaceutical companies found that 22% had segmented their supply chain. Interestingly, 49% of respondents said the ideal supply chain structure would be tailored to specific industry segments.

Pharmaceutical makers are embracing supply chain segmentation because the industry faces different challenges across the globe, Davis said. The U.S. market for medicines places an emphasis on high service levels, whereas Europe, because of its regulatory environment, demands more agility for rules compliance. And in emerging markets, where product buyers have limited incomes, pharmaceutical companies must figure out ways on how to deliver their products at the lowest supply chain costs.

Unlike 5 years ago when companies were talking about supply chain segmentation, consultant Daniel Swan of the firm McKinsey said more companies have begun executing that strategy, leveraging more mature supply chain capabilities in the face of competitive business pressures. Although he's found supply chain segmentation applied across diverse industries, Swan said that two industries have used this practice in particular. CPG companies have used supply chain segmentation to address stock keeping unit (SKU) complexity and the increasing diversity within their SKU base. High-tech companies have also engaged in segmentation as a way to address lead times and rapidly increasing innovation cycles.

Craig, who works with medium-sized companies on supply chain segmentation, said he's seen interest pickup in this strategy particularly in the past year. In his view, interest has grown on the part of mid-sized business as big companies in their markets have squeezed them. Chief executives at mid-sized companies are looking at segmentation as a way to tier their supply chains and to refocus their limited resources. "Market reality is setting in," said Craig. "They are trying to reassert themselves [against big companies]."

Although pharmaceuticals, high-tech and consumer products makers are the current leaders, segmentation may offer a way forward for retailers. Faced with the challenges of omnichannel commerce, retailers are juggling inventories and service levels for both online and brick-and-mortar sales. More than any other industry at the moment, retailers engaged in e-commerce need to segment their customer base, adjusting service levels based on the price that the buyer is willing to pay.

It appears that e-commerce leader Amazon.com, Inc., headquartered in Seattle, is already doing this to some extent. If a consumer opts to join the Amazon Prime program and pay the $79 annual fee, then Amazon provides 2-day shipping with no additional charge. Amazon Prime allows consumers to self-declare themselves to be "loyal, steady customers" worthy of special treatment.

The problem for e-retailers competing against Amazon is that many online buyers expect free shipping as part of the deal when they buy products over the Web. Of course, free shipping isn't "free" as delivery providers must be paid for their services. The decision to whom to provide free delivery service requires segmentation analysis, especially since retail executives with backgrounds in merchandising and marketing often lack detailed knowledge of distribution and transportation. "Let's face it, most retail executives don't come out of the supply chain world and struggle with understanding supply chain investment, costs and complexities," wrote Descartes Systems executive Chris Jones in a November 2013 blog for *DC Velocity* magazine titled "Don't Wait for the Inevitable in Omni-channel Retailing." "However, they do get margin erosion and especially that the inevitable shift to more on-line sales that they are pushing will drive lower margins."

Segmentation gives an omnichannel retailer an analytical method to ascertain whether the profits justify the expedited picking and delivery. It gives the retailer a way to determine whether it makes financial sense to provide every online customer with same-day delivery. (The topic of omnichannel commerce will be explored more in depth in a later chapter.) "The decision of how much to invest in enhanced services like omnichannel retailing and same-day deliveries should and must be determined by profit-based segmentation," wrote Byrnes on his blog for *DC Velocity* magazine in November 2013. "It is not a question of whether to offer these services, but rather to whom to offer them."

Segmentation for online selling may be the only way a web merchant can obtain decent profit margins. To do this, they'll have to use analytical software to determine break-even points for the various shipping methods to get merchandise in the hands of consumers. That may mean some consumers in certain regions such as rural areas have to pay a variable shipping fee to get merchandise bought online, while others—perhaps those living in metro areas with the household density to cover delivery costs—do not. Segmentation may even mean that the e-retailer may not offer certain products for sale online because they can't get enough of price margin to cover the fulfillment and delivery costs.

For retailers who want to both sell products online and in their physical stores, the question of appropriate service levels becomes even more challenging. In fact, segmentation strategy may be the only viable way for a retailer to serve both online and store channels and make money at it. "Omnichannel retailing is a very powerful marketing approach that is now emerging. Many companies see this as necessary to stay competitive. Yet it can be enormously expensive and unproductive if not constructed in a thoughtful way," Byrnes blogged. "The decision of how much to invest in enhanced services like omnichannel retailing and same-day deliveries should and must be determined by profit-based segmentation."

As retailers wrestle with how to structure their supply chain operations to serve both traditional in-store and nontraditional online shoppers, they will need to understand their operating margins for products, channels, and customers. "Profit-based segmentation is absolutely critical for retailers, both for traditional business, and for omnichannel and possibly Sunday deliveries," said Byrnes. "In virtually all companies, the majority of transactions and customers are largely unprofitable, while the minority are highly profitable. Why give unprofitable customers enhanced services that make them much more unprofitable?"

In a word, supply chain segmentation gives omnichannel retailers an approach to securing profit margins in the face of market pressures for expanded service levels. But this situation isn't unique to retailers and consumer goods manufacturers. Every business faces intense pressures for heightened services in a global economy that's digitally connected.

Although segmentation promises lower costs and higher revenues, it does require companies to run multiple supply chains. That would have been mind-boggling a decade ago. Just imagine giving each customer different criteria for procurement, inventory planning and replenishment, transportation, order management, warehousing, and customer service.

Fortunately, today software makes the practice of customer differentiation less of a hassle. That's because multiple supply chains required for segmentation are all virtual. Business rules can be set up in a software program that keeps track of preferential customer treatment. Rules can be defined for each customer segment. Rules can be put in the warehouse management systems that allocate inventory differently. Rules can be put into the transportation management systems that specify different delivery criteria for different customers. The software knows that a top overseas customer is allowed to get his products delivered by higher-priced airfreight, while a poor customer only gets delivery with slower-going, lower-priced ocean shipments.

Although software makes the practice of supply chain segmentation easier to do, many supply chain managers still remain hesitant. Logistics Bureau managing director O'Byrne, whose firm has offices in Australia and Asia, said reasons for that hesitancy stemmed from the fact that not enough supply chain executives understand how to do segmentation or appreciate its value. In line with that, he said that supply chain executives "perceive they have more pressing issues or lack the resources or systems to do" segmentation.

Gartner analyst Tim Payne told me that many companies are still wrestling with how to put a segmentation strategy into action. The process of supply chain segmentation has three phases that must be done in turn—design, planning, and, finally, execution, he said. Although more companies have started work on the first two phases, they have not figured out how to align segmentation with execution events such as customer orders or supplier delays. "I find most companies that are doing segmentation haven't even begun to think about how they will enact out their segmentation strategy," he told me. "There is a higher penetration for the design and plan phases of segmentation."

Another consultant indicated to me that when faced with the nitty-gritty work demanded for a segmentation strategy, many companies retreat. After companies conduct some pilots, they run out of energy and ideas when it comes to implementing "the more complex and subtle actions," said consultant John Sewell with the London-based firm Crimson and Co. "Everybody has some obvious low variability and high volume products and high variability and low volume products," he said. "The challenge is what to do with all the stuff in between."

One reason for lack of action may be that companies, especially mid-size ones, haven't put into place the financial measures that lay the basis for segmentation awareness. The Council of Supply Chain Management Professionals and National Center for Middle Market did a joint study in 2013 that surveyed some 200 financial decision makers across the supply chain in the middle market companies on the financial metrics used for margin management or the ability to understand where money is made or lost in the business. By the way, middle market companies were defined as those businesses with annual revenues between $10 million and $1 billion. The report made the claim that while the middle market represents only 3% of all U.S. companies, that portion accounts for a third of all private-sector jobs.

What the survey found was that the decision makers, whether working for raw material suppliers, manufacturers, wholesalers, distributors, retailers, or service providers, still used the most basic financial measurement—profit margin. Only a few used more sophisticated metrics. Ninety-six percent of the survey respondents used profit margin as a financial tool. While 64% used the financial metric profit by product and 61% used profit by customer, only 14 used cost to serve. The study also found that in guiding supply chain decisions, profit by product was the metric most often used, while cost to serve was at the bottom of the list. "While not all products, services or customers will be profitable for your company all the time, it only makes sense to try to understand where you are making money and losing money in the business," the report stated in its conclusion. "Operating in the dark and simply hoping to win more often than lose is no way to run a business."

Supply chain segmentation illuminates the underlying components of the business. Unless companies have segmented their supply chains and understand how costs align with customer and product classifications, they will make blind, in-the-dark decisions on how to use their capabilities. "A supply chain that says yes to every customer without enforcing required trade-offs will erode profitability and cannot manage a set of standard, repeatable processes. It overservices certain customers while inconsistently underserving others," Davis told me. "A supply chain that says 'no' to any customer request outside the standard process is stuck with one-size-fits-all processes and cannot satisfy varying customer needs or pursue new profitable growth opportunities."

As companies seek to run their supply chains on the basis of demand signals, they'll be required to make quicker decisions about what to make and who to supply. A supply chain segmentation strategy provides a playbook for their responses. Unless a company has segmented its supply chain when it becomes demand-driven, Byrnes told me, then the company will end up overpricing and underserving its best customers while at the same time underpricing and overservicing its worst ones.

Despite the trend toward more demand-driven supply chains, discussed earlier in the book, only forward-thinking companies have undertaken the entire process of supply chain segmentation. Davis has estimated that 15% or less of companies have segmented their supply chain.

That said, running different supply chains for different customers is not a business option anymore. It's a necessity. What's changed is the need for periodic reassessment. As customers change, products change, and markets change, companies will have to regularly reassess the segmentation of the supply chain. The business landscape has become too unpredictable to segment a supply chain even once a year and to assume that the resulting segmentation breakdown will remain correct for an extended period of time.

Consequently, protean supply chains will have to revise segmentation breakdowns in sync with fluctuations in the marketplace, readjusting their supply chain service policies to changes in customer and product classification. Companies will have to build software models that can do segmentation updates quickly. If companies want to stay profitable in a volatile global economy, they'll have no choice.

BIBLIOGRAPHY

Jonathan L. S. Byrnes. Join the finance revolution. *Supply Chain Quarterly*; Quarter 4, 2010.

Jonathan L. S. Byrnes. Profit-based segmentation, omnichannel retailing and same-day deliveries. *DC Velocity* blog; posted online November 21, 2013.

Matthew Davis. Gartner supply chain top 25 and supply chain segmentation. Blog posted online April 1, 2013.

Goetz Erhardt. Aligning customer segmentation with industry realities to achieve high performance in the chemical industry. Accenture white paper; 2011.

Janet Godsell. Thriving in a turbulent world: The power of supply chain segmentation. Cranfield University School of Management; August 2012.

Chris Jones. Don't wait for the inevitable in omni-channel retailing. *DC Velocity* blog; posted online October 21, 2013.

Douglas M. Lambert. Which customers are most profitable? *Supply Chain Quarterly*; Quarter 4, 2008.

Life sciences supply chain benchmarking survey 2013. World Business Research. 2013.

Llamasoft. Cost-to-serve optimization. White paper; October 20, 2010.

Margin management among U.S. middle market firms. The National Center for the Middle Market and the Council of Supply Chain Management Professionals; 2013.

Rob O'Byrne. Cost to serve video. YouTube; October 18, 2008.

Tim Payne. Hype cycle for supply chain planning, 2013. *Gartner*; November 22, 2013.

Paul W. Schroder and David M. Powell. Rules of engagement: A better way to interact with suppliers. *Supply Chain Quarterly*; Quarter 4, 2012.

David Simchi-Levi, Annette Clayton, and Bruce Raven. When one size does not fit all. *MIT Sloan Management Review*; 54(2), 14–17, 2013.

Daniel Swan, Sanjay Pal, and Matt Lippert. Finding the perfect fit. *Supply Chain Quarterly*; Quarter 4, 2009.

Daniel Swan, Matt Lippert, and Nitin Chaturvedi. Turning your supply chain into a competitive weapon through segmentation. McKinsey & Co. presentation; March 4, 2012.

Kelly Thomas. Supply chain segmentation: 10 steps to greater profits. *Supply Chain Quarterly*; Quarter 1, 2012.

Donna Wharton. Segmenting your supply chain for success. *Supply Chain Quarterly* webcast; April 2012.

CHAPTER 6

THE OMNICHANNEL CHALLENGE

Online sales continue to grow, posing a challenge for traditional retailers. In a report released in March 2013, Forrester Research stated that e-commerce generated $231 billion in 2012, accounting for 8% of overall U.S. retail sales. The research firm said e-commerce should reach $370 billion by 2017 as more consumers shop for goods with their mobile devices.

Another firm, Frost & Sullivan, has predicted that global online retail sales will reach $4.30 trillion by 2025. In its study, "Global Mega Trends and Their Impact on Urban Logistics," Frost & Sullivan said, by 2025, nearly 20% of retail purchases will result from online channels. Indeed in leading markets such as the United States and the United Kingdom, the firm said nearly 25% of retail sales will come from online orders.

The rise of online retailing is coming at a time when retailers confront a lackluster economy and changing demographics. As the population ages in Western countries such as the United States, retailers can expect elders living on fixed incomes to have fewer discretionary dollars and to be more predisposed toward spending those dollars on health services than luxury goods. While online sales

Protean Supply Chains: Ten Dynamics of Supply and Demand Alignment, First Edition.
James A. Cooke.
© 2014 John Wiley & Sons, Inc. Published 2014 by John Wiley & Sons, Inc.

may be growing, traditional store sales are likely to stay flat at best. That same situation that exists in the United States exists on the other side of the Atlantic, in Great Britain. In an October 2013 presentation, "Omnichannel Fulfillment . . . Key Strategies and Tactics," Stuart Higgins with the British firm LCP Consulting Ltd. said that individual retailers in United Kingdom could see store sales range from a 2% increase to a 5% decline, while online and multichannel sales grow about 15%.

That leaves the retailers chasing a younger, more tech-savvy generation of buyers. Smartphones and tablets are creating a new mindset on the part of young shoppers, who expect product availability at the lowest price. They'll visit a store only to see and touch the sought-after item—a practice dubbed "show rooming." And instead of buying the item from the store, they'll order it online from a digital merchant, using a smartphone.

Regardless of whether the item gets bought online or in a store, there's now a deep-seated expectation on the part of the consumers that the item will be available anywhere anytime. And that means retailers must have stock on hand for both sales channels: the store located online and the store located in the shopping mall. It's no wonder that the competition for e-commerce between online merchants like Amazon.com, Inc. and traditional retailers will become intense in the decade ahead.

At first blush, the simplest supply chain solution would be for retailers to view all their inventory, be it in a store or a warehouse location, as a single common pool to draw from for all types of sales. According to Kurt Salmon, a global management and strategy consulting firm with a large practice in the retail industry, store chains such as Ann Inc., Bed Bath Beyond, Jones Group, Macy's, Polo-Ralph Lauren, Toys "R" Us, Urban Outfitters, and Walmart are all taking items out of some stores to ship web-placed orders. In an October 2013 press release announcing that it was opening two more distribution centers (DCs) dedicated to filling online orders, Wal-Mart said that more than 10% of the units ordered on Walmart.com were being shipped to a customer's door from a store.

And more retailers are planning to join companies like Walmart in having online orders filled with products taken from the store shelf. A 2013 research on omnichannel commerce conducted jointly by ARC Research and *DC Velocity* magazine found that 35% of 177 retailers in the study took merchandise out of their stores for online orders and another 18% were filling orders from select stores. And that practice is just catching on, the study found; some 56% of surveyed retailers not now filling online orders from store stock planned to adopt that approach within the next few years.

To assist retailers who want a common inventory pool, software vendors offer an application that makes it possible to see all the inventory held across locations, whether in the central DC, regional depot, or the store. It's called a distributed order management (DOM) system, and this application generally sits over the enterprise resource planning system, warehouse management system, and the store's own system.

A number of retailers have begun using this software, as the DOM system determines the best location to fill and ship an order from. If the merchandise desired by an online consumer rests in the store backroom or sits on the store shelf, then the DOM tells can direct the store to package and ship the item rather than having it done in a DC. The DOM can even ask the online buyer, when the order is placed, if he or she wants to pick it up at the store location.

If the retailer's DOM has visibility into a supplier's warehouse, then the store merchant can even have the supplier do the job of pulling and shipping the wanted item. That's a big advantage to a merchant in that the retailer does not have to hold on its books or bear the costs of keeping an odd-size or unusual item in stock. The merchant can offer on its website a huge assortment of products, but then leave it to the supplier to warehouse and fill the order. By the way, Amazon in its early days as an online book merchant did just that—had a small publisher ship a web-ordered novel direct to the buyer.

For retailers, the advantage of a common inventory pool is that they are not stockpiling excess inventory. And all items are part of the pool. From the retailer's perspective, it makes sense to take a slow-moving item on a retail shelf, use that slow mover to fill an online order, and not have to mark down the price of the good to entice a consumer into making a purchase. It also means a retailer can offer a customer more variety than what's possible in the limited confines of a physical store. "The strategic goal is to satisfy clients," said Kerry W. Coin, the founding principal of the firm The Kerma Group and a former chief logistics officer at ANN Inc. during a discussion on omnichannel at the 2013 CSCMP Annual Conference. "You buy 10,000 items but the store has a limited assortment of those. It can only carry so many colors and sizes."

On the other hand, DOM is not an inexpensive solution; in fact, it can cost up to million dollars to install and integrate that application with other pieces of supply chain software such as warehouse management systems, order management systems, front-end e-commerce systems, and customer relationship management systems. Although DOM systems look appealing as a solution to the challenge of filling both physical store and online orders, it's far from perfect. Although store computer systems inform the DOM when the merchandise has arrived at the retail outlet and when merchandise gets sold, so it can update the master list of inventory holdings in the common pool, it often lacks up-to-the-minute visibility in the time period between a product's entry and exit in a store. And that raises potential conflicts over item allocation. It's possible for a store shopper to pick up an item of merchandise on the shelf that's been flagged for an online order. It's possible that a shopper has picked up the item and then moved into another location. In addition, because stockers in stores don't generally follow the standard put-away practice of a DC, which involves scanning the item bar code and then storage location bar code to create a location match in the computer system, stores don't often know the precise location for specific items on a shelf. The store lacks slot-level visibility.

Because of the possibility of inventory mismatches and a lack of up-to-the-minute visibility of an item's location, DOM software often sets a threshold target for ceasing online inventory fulfillment from the store. For example, a store received a shipment of 100 sweaters. As the stock of sweaters drops either because shoppers buy the sweaters or the merchandise is picked for online orders, the DOM keeps close tabs on the amount of store inventory. Once the system determines that the store has only 10 sweaters left in stock, it will not pick from the store because the application can't be certain that items are actually there.

Indeed, poor in-store real-time slot visibility is one of the challenges facing most retailers trying to pick online orders from the store shelf. The 2013 study by ARC Research and *DC Velocity*—alluded to earlier—found that only 30% of 177 retailers canvased said that inventory accuracy of their stores was 98% or higher. As a point of comparison, most DCs today can track inventory at 99% or higher, thanks to the use of warehouse management software and automatic identification equipment. Warehouse management software knows the exact bin or slot location of products and parts in storage. Hence, the software can direct a warehouse worker to the precise location to pick an order. That's not the case for most retailers engaged in omnichannel commerce. They don't have slot-level accuracy of where the merchandise lies in the store whether in the backroom or upfront.

But there are operational issues with stores shipping web merchandise. To start with, many retailers engaged in omnichannel selling don't understand the costs associated with picking orders in a store. Although 78% of the respondents in the ARC/*DC Velocity* research study knew the costs for picking individual items in a DC used for e-commerce, only 38% knew their costs associated with picking from shelves in the front of the store. How can a retailer be profitable if it doesn't understand its costs for store fulfillment?

For retailers eyeing an omnichannel strategy that uses stock in the store to satisfy an order placed on the web, the pick-pack-and-ship from a store requires store associates to be adept at multitasking. Store personnel will have to be pulled away from duties like manning the cash register, stocking shelves, or assisting a shopper. Cashiers have to be cross-trained to be order pickers. On top of that, retail store managers have typically not allocated additional manpower to this task in the store budget.

Kurt Salmon consultant John Seidl has seen firsthand just how challenging store fulfillment can be for retailers. Seidl has worked with numerous retailers in this area and he told me a story about one of his clients. During the 2012 holiday season, a retailer had its stores handle 17% of its online order fulfillment. Yet 64% of its returns and calls to customer service that holiday season resulted from orders shipped out of its stores. "They [retailers] throw a folding table in the back room of the store and put a printer on it to print labels and say—go for it," he said. "The stores are not designed for it. The work space is not appropriate. The workforce is not trained. And it creates chaos."

A store by function is designed to serve shoppers and not function as a DC. So if retailers want to be effective with shipping from their stores, they will be forced to establish a special staff within each store outlet to handle just e-commerce fulfillment. The British grocer and retailer Tesco PLC, for example, has used dedicated personal shoppers to select products for its online customers in its stores since 1997. In the U.S. department store chain Sears is also using dedicated labor to pick merchandise from shelves for online orders in the limited stores the retailer has designated for pick-pack-and-ship. The special staff reports to the store manager.

In my view, this cadre of distribution workers within the store should report to the head of supply chain operations rather than the store manager. This special "web-order" team should not get involved in typical retail duties but should rather focus on their mission of locating merchandise for servicing online customers. Like staff in a DC staff, the web-order team will have to be outfitted with voice technology to direct them to the correct store location for item selection and they'll have to carry handheld radio-frequency scanners to record their activities. Although hard-pressed retailers may be resistant to additional store personnel with a sole mission, they may have little choice if they want to engage in omnichannel strategy. In fact, retailers putting in self-checkout stations to reduce headcount may simply end up reassigning store labor rather than replacing employees. Cashiers replaced by checkout automation may have to be reassigned as in-store pickers.

In handling online orders in-store order selectors may have to take back tasks retailers had relegated to suppliers. In the past decade, in a bid to lower their own distribution costs, many retailers and manufacturers have put back on their suppliers some of the more onerous work of prepping merchandise. Take a shirt, for example. If it's intended for sale in a retail store, then the supplier will affix a price tag and size strip as well as fold the article of clothing such that it fits neatly on a store shelf. On the other hand, a shirt intended for online sales is prepped with collar stiffeners and pins for maintaining its proper shape while making the journey on a delivery truck to the consumer.

So, if a retailer wants to maintain a single pool of inventory for an "omnichannel" selling strategy, then it has to have its workforce handle article preparation whether in the store or at the DC. Given these considerations, it should come as no surprise that some retailers still prefer a "multichannel" strategy—they have set aside one DC to specifically handle online stores and another one to replenish stores. In fact, when e-commerce got under way in the 1990s, most retailers opted for separate DCs for online orders and traditional stores.

Christine Fotteler, a vice president for retail solution management and supply chain for software giant, SAP, said some retailers in Europe and United States are still running one DC for web sales, another for traditional stores. "One reason is because the warehouse for online [single item picking] is organized differently than the warehouse for bulk picking," she said. "Sometimes these are different areas in the same DC but sometimes it's an entirely different DC. Several companies outsource their web business and they handle the entire web inventory separate from their brick-and-mortar inventory with no global view of inventory. Most companies want to get away from that practice but not all have managed."

The complication is that work processes differ in DC that supports store replenishment from those that service web orders. In an operation supporting brick-and-mortar stores, trucks generally bring in product in cases loaded on pallets or skids. The pallets are placed direct into storage in the warehouse or the pallet load broken down such that the cases are placed in building storage. When the store needs to be resupplied, then the retail distribution removes pallets or cases from storage racks and loads them into a trailer or truck for the replenishment run to a store.

Some retail DCs perform what's called cross-dock operations. Walmart has a reputation for being excellent at cross-docking merchandise from suppliers into its stores. This ability to cross dock allowed Walmart in the 1990s to reduce logistic costs and provide a higher level of product availability at its retail outlets. In cross docking, trucks bring in goods on pallets into a DC. The pallets, usually carrying a single make of product, are off-loaded quickly from each truck. The pallets from several trucks are then combined to make up a shipment of assorted product that another truck takes to a specific store. In some situations, workers in the DC will pull apart the cases on a pallet and then combine them with cases from another pallet to create a mixed-case or rainbow pallet that holds a variety of products (say, different kinds of cereal boxes or soda). The rainbow pallet then gets shipped to a specific store. In a cross-dock operation, there's no storage—the warehouse acts as a transfer point.

But filling online orders requires picking, packing, and shipping single items. Lots of items. Workers have to open cases and pull out each item—either before the cases are put away into storage or more often when the cases come out of storage. Direct-to-consumer fulfillment for e-commerce is a very labor-intensive warehouse operation because it means handling individual orders or "eaches." That's why many experts prefer the term "fulfillment centers" for facilities that handle individual orders as opposed to DCs, which handle pallets and cases.

The hours of operation at a DC to support the channel of store replenishment are less extensive generally than that of fulfillment center supporting the channel for online retailing. Online orders come in from shoppers in multiple time zones at all hours of the day or night. Unlike a traditional warehouse, e-commerce operations tend to run 7 days a week with at least two shifts per day to fill customer web requests. And it's not uncommon for many of these operations to run 24 hours a day.

Large retailers have become quite efficient at moving cases, pallets, and garments on hangers through a warehouse rapidly and efficiently. But if the warehouse worker has to open the case, pull out an item like a shirt, place in a box, well, that entails more work even if no prepping is required. Automation isn't really practical as it's just not easy to mechanize the task of opening boxes and pulling items out of them. Although it's possible to use robotic mechanisms if the carton size and shapes are always the same and the objects to be removed the same, there's too much order variety in most DCs doing e-commerce.

Order selection consumes much of the work activity in a DC. In general, picking constitutes between 55% and 75% of the labor costs in a distribution operation, said Helgi Thor Leja on an October 2013 *DC Velocity* webcast on same-day fulfillment. Leja is a leader in the industrial and electrical distribution practice for the consulting firm Fortna.

To get more work out of their workers, many big DCs have turned to software such as workforce management and labor management in the past decade. Labor management allows a company to measure and monitor worker activity in performing such activities as receiving, put-away, and picking. The software relies on the recording of workplace activities as the basis for task comparisons. The recordings occur in conjunction with the use of bar codes and scanners. Whenever a worker reads a bar code with a laser scanner, it creates a record of the activity. When a case of product gets off-loaded from a truck and then gets scanned to verify the contents of an inbound shipment, a computer system makes a notation of task completion. When a warehouse worker places a product in a bin in a storage rack and then scans a bar code on the rack so a computer system knows the inventory location, that creates another activity record. All those of activities—picking, putting, and shipping products—get "time-stamped" by a computer system.

All those time-stamped records make it possible to measure worker performance. Because the software is collecting all those time stamps, a "labor management system" application can compare an individual's work activity against a preset standard for that activity. Often a company must bring in an industrial engineer to analyze the tasks being performed in the warehouse, conducting time–motion studies to determine the correct amount of time required to do a specific task, to set those standards. Some software programs, by the way, do contain preset standards for tasks, based on data that the vendor has collected from its client base. Managers can then use these labor standards to ensure that workers meet a high degree of consistent performance. As companies have looked to optimize warehouse operations, labor management systems have proven effective. By the way, although labor management systems were originally developed as separate stand-alone application, most warehouse management systems today from large software vendors offer this component.

Along with wider use of labor management systems to bring about workforce efficiency in each picking, companies have also embraced workforce management software. That application deals with scheduling, tracking, and planning labor. This particular type of software automates the chore of assigning workers, taking into account such variables as individual availability, overtime status, and skill sets. (Is the laborer certified to drive a forklift truck?) This software can make sure that part-time workers don't exceed the hour quotas, which would mandate paying benefits like health care. For e-commerce fulfillment, whose orders spike during the holiday season, this software has helped companies handle the task of managing seasonal labor and ensuring adequate staff of qualified personnel on hand.

Although those types of software certainly help to improve the productivity of the warehouse workforce, it does not lessen the manual work that has to be done for filling individual orders. Also keep in mind that an omnichannel DC must perform dual functions: send cases or pallets to a store and also pick individual orders.

It should be noted that some retailers, especially those that started out as catalog merchants, do have experience in filling individual orders or eaches. They are better prepared for omnichannel commerce because they have already made provisions in their warehouse for handling eaches, even if the individual order selection has only been for store replenishment. The warehouse space has been compartmentalized with areas designated for particular activities. One area in the warehouse is set aside for workers to fill individual orders, another to handle store replenishment. Mezzanines work well for this purpose.

The task of omnichannel distribution would be a lot easier if the entire warehouse could be automated with a turnkey solution. Although some material handling equipment makers are trying to develop this type of solution, at the moment, the best a company can do is to bring together an array of equipment to support both the picking of eaches and the handling of traditional case and pallet replenishment under a single warehouse roof.

To this end, many DCs employ computer-controlled equipment with high-speed sortation that remove cases from warehouse storage and bring them down designated conveyor lines to workers who open those cases and remove the ordered item. The workers then place the picked items into a carton, tote, or tray. When filled, the totes or cartons are next placed on a mechanized or roller conveyor system for travel to a packing station in the warehouse. There the goods get boxed, labeled, and sealed. The packaged goods are often then sent on another journey on another conveyor to a designated warehouse door for pickup by a trucking company or parcel carrier.

The use of conveyors and automated equipment helps to reduce the need for workers to walk around the building. In the article "Material Handling Equipment for Multichannel Success" (*Supply Chain Quarterly*; Quarter 4, 2013), consultant Matt Kulp pointed out that a reduction in travel time becomes an important consideration for a warehouse handling direct-to-consumer fulfillment. That's why many warehouses assign workers to "pick zones" for selecting items only within a short reach of them. "One traditional approach to reducing travel is to build pick modules that give the picker access on foot to SKUs three or four levels high," wrote Kulp, a director and principal with the St. Onge Co., headquartered in York, Pennsylvania. "Picker travel is limited to within their pick zones."

To cut down on worker travel, DCs engaged in pick-pack-and-ship also use a parts-to-person or goods-to-person material handling approach. Mechanized conveyance vehicles are one way to do this; the vehicles provide a way to take a tote, a box, or a set of garments on hangers from storage to the worker. One of the leading providers of mechanized vehicles was Kiva Systems Inc., and many experts believe Amazon purchased Kiva in March of 2012 as a way for the e-commerce leader to secure a competitive advantage in the area of goods-to-person technology. Kiva made orange-colored, lozenge-shaped "bots" that bring storage racks to workers for picking and packing. By the way, although Kiva is the best-known maker of robotized conveyance vehicles, there are other such machines on the market. The argument in favor of robotized conveyance is the same as the one for setting up pick zones. It's more time efficient to bring the goods to the worker doing the packing than have the person go roaming through the warehouse to find the item.

There are other ways to accomplish a "goods-to-person" approach in a warehouse besides robotized conveyance vehicles. Another way to do this involves the use of automated storage and retrieval equipment that use shuttles. Computer-controlled machines use cranes and conveyance vehicles to take pallets, totes, or bins from storage racks and shuttle them to a station at the end of an aisle.

Even though a machine can eliminate the time required to fetch an object, a warehouse worker still has to break open the case and pull out a single item. And that's the underlying problem. You can automate the process of item movement but not item extraction. The best way to open the box—at least for now—is with a pair of human hands. It doesn't matter at what point the box opening occurs. It doesn't matter if the case is opened when the goods arrive at the warehouse so single items can be placed into storage. It doesn't matter if the case is opened when pulled out from storage—the normal practice. At some point, the box has to be opened to take out a single item to fill an order.

There's no easy way to automate case opening if the size and shape of items vary from one box to another. A machine can be designed to remove milk jugs if that's the only type of item to be handled in a warehouse. But that's generally not the situation in most distribution or fulfillment centers. They handle a variety of products. One corrugated box may contain tools like hammers, and the next box may hold clothing.

The answer would be a robot. Or more precisely, a humanoid robot. That's because a humanoid robot could use its hands to pry open the box and remove the item. And the humanoid robot would have eyes to distinguish a fleecy sweater from a leather jacket and then use an arm with fingers to delicately pull out the item.

But the robots in current used today in warehousing and manufacturing don't have the prehensile grasp found in humans and chimpanzees. Unlike what's described in science fiction lore, today's industrial robots are really machines that perform automated tasks. Even though no real robot exists that can solve the problem of picking eaches yet, there are a number of interesting development under way that are promising. For example, a company called Rethink Robotics Inc. in Boston, Massachusetts, is working on a possible answer. The company has developed a robot called Baxter, which has a camera face and two mechanical arms. If Baxter or another robot like it can be developed with fine motor skills, then it would solve the problem of item extraction. But the robot would have to be able to pick up delicate clothing from a box as well as pick up solid items of different sizes and shapes.

Until such time as robots become available for picking and packing eaches, order fulfillment for single orders will remain primarily a manual task. No doubt, that's why a company like Amazon in 2013 went on a hiring spree to staff all the warehouses it planned to open as part of its ongoing efforts to dominate e-tailing. *DC Velocity* magazine reported that the e-commerce giant planned to employ another 5000 workers for 17 fulfillment centers across the United States and another 2000 full- and part-time customer service representatives.

Amazon will likely need all those extra hands in its warehouses. That's because order fulfillment is going to become even more demanding in the years ahead to meet customer expectations for same-day delivery, whether the e-commerce is business-to-business or business-to-consumer. "My prediction of the new norm is that 100 percent of orders in the future will soon require same-day fulfillment and if not 100 percent it's going to be very close," said Leja on a *DC Velocity* webcast titled "Same-Day Fulfillment: It's a Balancing Act. "In order to meet growing customer expectations for shorter order delivery times, companies are driving distribution to reduce order fulfillment from hours to minutes."

Along with fast picking in the fulfilment center, same-day delivery involves bringing the product to the buyer's door. The transportation component could prove even more challenging than the picking, packing, and shipping. For starters, there's the issue of delivery costs. In a December 12, 2012, *New York Times* article about Amazon's foray into same-day delivery, Yossi Sheffi, director of the MIT Center for Transportation and Logistics, wondered about the high cost for providing this heightened level of service. He estimated that it would cost $10 an order to fill up truck and one-off delivery would cost up to $50 alone.

The leader in e-commerce, Amazon is not shying away from the challenge. In 2013, Amazon was testing a service called Amazon Fresh in select zip codes in Los Angeles and Seattle. A customer places an order online for grocery-store fare: beverages, breads, canned foods, dairy, cheese and eggs, deli meats, frozen foods, meat, produce, and seafood as well as pet, health, beauty, and household and cleaning products. James A. Tompkins, chief executive officer of the consulting firm Tompkins International in Raleigh, North Carolina, told me that in Los Angeles, Amazon Fresh is leasing drivers and vehicles from Ryder, a company that provides fleet management services. "The key to making this work is delivery density," said Tompkins. "Amazon is controlling this by going to densely populated areas and limiting the zip code coverage for delivery. Amazon is doing well on this."

Although it's unclear what Amazon's long-range plans are for this service, the e-commerce giant has raised the stakes. It should be noted that Amazon still continues to use parcel carriers to deliver online merchandise. And in the fall of 2013 Amazon announced a partnership with the U.S. Postal Service (USPS) to deliver packages on Sunday in the Los Angeles and New York metropolitan areas. But those two cities are just the start of the Sunday-delivery program. In its press announcement, Amazon said it and USPS would expand the program in 2014 to include such cities as Dallas, Houston, New Orleans, and Phoenix. "If you're an Amazon Prime member, you can order a backpack for your child on Friday and be packing it for them Sunday night," said Dave Clark, Amazon's vice president of worldwide operations and customer service in the announcement.

By piloting both Sunday and same-day delivery, Amazon has upped the ante for direct-to-consumer shipping. It has raised the stakes for all retailers or distributors engaged in e-commerce as they will be expected to match those service levels at least in those areas where Amazon is already providing heightened delivery. Products will have to be now picked within minutes of a receipt of an online purchase. That will force companies to scrutinize every single aspect of their distribution or fulfillment center operation to find ways to boost speed of handling and throughput.

Not to be outdone by its rival, Walmart is also experimenting with its own same-day delivery service called "Walmart to Go." In 2012, Walmart launched a same-day home delivery service for general merchandise in five cities, including Denver, and later, a same-day grocery service in San Jose and San Francisco. In October 2013, Walmart added grocery deliveries to Denver.

Participation in the Denver trial was limited to folks who had requested an invite on the company's website and then gotten approval to take part. Test participants signed up for a special Walmart to Go account. They can then order tens of thousands of items from fresh produce to dairy, from meat to baked goods, from health and beauty aids to toys and electronics. Items carry the same prices as the local Walmart Store. Customers can select 2-hour slots (noon to 2 p.m., 2–4 p.m., etc.) for delivery of their online ordered items. Same-day delivery is provided if orders are placed by 8 a.m. Mountain Time.

Walmart fills the online orders from local supercenters in the Denver area. What's interesting is the picking process that's employed. In a video for journalists on the Walmart website, order selectors can be seen picking the merchandise in the store the same way it's done in a DC. The order selectors use bar-code scanners to confirm the picks before placing the items in a tote. "We think of our stores as forward deployed fulfillment centers," said Walmart spokesman Ravi Jariwala. "They allow us to provide additional value to customers."

The online-ordered items are brought to the buyer's home on green trucks with the Walmart to Go name emblazoned on the side panels. The trucks are equipped with separate frozen, refrigerated, and ambient compartments. A Walmart spokesman said that Penske Logistics is operating the trucks used in the online grocery test in Denver. Although some Denver test participants got free delivery as an introductory offer, Walmart is testing delivery charges in the $5–$10 range.

While Amazon and Walmart are pioneering the use of in-house delivery fleets, at the moment, other retailers engaged in selling online products appear to be relying on parcel providers for delivery. In the omnichannel study mentioned earlier, 80% of the retailer respondents relied on parcel deliverers to some extent for this task. The study also found stores were doing delivery with couriers, the in-house transportation fleet, and even in some cases store personnel. Store staff drove their own vehicles, took the subway, or hoofed it on foot to make those deliveries.

Retailers are also experimenting with so-called lifestyle couriers—individuals who transport packages in their own vehicles. In Europe, transport companies are signing up and screening individuals who want to make extra money doing these kinds of deliveries. *DC Velocity* magazine has reported that the U.K. delivery company Hermes has 7500 lifestyle couriers on call and DHL has launched a service in Sweden—MyWays—to recruit citizen drivers.

Although Ma & Pa's Pizza stores in the United States often make deliveries with individuals using their own cars, there are lots of complications here. There are questions of liability for using a personal vehicle on a commercial trip. There are security issues for both the homeowner and driver. How many pizza delivery drivers have been robbed? And there are performance issues since the customer expects prompt delivery of an item, intact, and in working condition.

To avoid complications and to be efficient while meeting time commitments, retailers may have to set up their own dedicated team of delivery drivers using company vehicles. In this regard, Wal-Mart Stores does have an advantage over Amazon and other retailers. That's because Walmart has considerable experience in truck operations, having run one of the largest private fleets in the United States for years. After U.S. transportation deregulation in the 1980s, when other companies dumped their in-house delivery fleet in favor of using for-hire truckers, Walmart opted to keep and expand its use of private carriage to supply its stores. Walmart's fleet of large trucks is dedicated to store distribution. Still, that truck-fleet operation provides Walmart with a basis of knowledge on seemingly mundane but important issues such as equipment maintenance and driver scheduling, knowledge that it can apply to the operation of a fleet of van vehicles for home delivery.

Make no mistake—same-day delivery will be a challenge even if a company operates it own vehicle fleet and has the population density within a delivery area in close proximity to a store or fulfillment center to justify the expense. That's because retailers will have to take on and master another task—transportation—that's not a core activity. Because of that same-day delivery, operations may be ripe for outsourcing. That may well open up a huge new market for third-party logistics, which already run trucking operations specifically for companies, a practice called dedicated contract carriage that's generally been used for hauling products from factories or DCs in big Class 8-size trucks.

Another key point in running a home delivery fleet from each store is the need for dispatch. Someone onsite at the store has to be responsible for managing and directing the trucks to the customer's address. In his June 2013 blog titled "The Last Mile Is the Last Word" on the *DC Velocity* website, Chris Jones argued that retailers will have to take full ownership of what's involved in home-delivery services to be successful. For retailers, that means "you need to have proof-of-delivery and post-delivery processes and supporting technology to ensure the job you expected gets done and to the level you expected," wrote Jones, an executive vice president for marketing and services at the software vendor Descartes Systems Group.

The handoff between the fulfillment center and the home delivery service will also have to be managed closely to avoid fumbles. "Retailers need to integrate their fulfillment systems with the delivery agent and the ownership of the delivery process needs to be seamless," said Jones in his blog posting. "There should have been clear instructions on the installation as part of the delivery manifest and then, when it doesn't happen correctly, provide feedback on what to go back and complete."

If retailers opt not to run their own delivery fleet from the store, they will likely have to turn to local couriers. In many respects, local couriers may be more suited to handling home delivery than the existing large parcel companies. Jim Serstad, managing director of Asia for Tompkins International, takes that view. Serstad has contended that local couriers are in the position to master navigation required to serve a local market. In China, where Serstad is based, the largest e-commerce companies have their own distribution networks, staffed by bicycle couriers. "It's amazing how fast things are done with e-commerce in China. I order and receive things in 4 hours in Shanghai," he said during a December 2013 conference call hosted by investment bank Stifel Nicolaus. "UPS and Fedex may not be able to do that. They can get things across the country but inside of a city may be a different game."

Because of the demand for higher service levels required for same-day fulfillment and delivery, retailers will need to employ a customer segmentation strategy as well. At the moment, online shoppers have grown accustomed to paying little or nothing for the items they order. But faced with intense price competition and shrinking profit margins, retailers may not be able to absorb the added expense for a same-day delivery, whether using couriers, parcel carriers, or an in-house team of drivers. "As more retailers adopt omnichannel retailing and move to a greater percentage of online business, their margins will actually decline," warned Jones in an October 2013 blog titled "Don't Wait for the Inevitable in Omni-Channel Retailing."

One way to address this margin pressure may be for retailers to charge fees for differentiated levels of delivery service. In other words, if a customer wants the product on the same day, he or she may have to be willing to pay extra, just as some pizza shops tack on a delivery charge for the driver coming to the customer's residence.

One British retailer reportedly has already started doing just that. It uses special software to determine the delivery costs and on the basis of those cost calculations provides the customer with delivery options that vary in price according to the time window. "With time becoming the new currency for many busy consumers, free may only need apply to "standard" all day home delivery windows, but not necessarily for deliveries at noon or 6 p.m., in 1-hour time windows or same day, " wrote Jones in a January 2013 blog at *DC Velocity* titled "Home Delivery Success Starts at the Order."

Online retailers will have to understand whether their customer places any monetary value on time. They'll also have to deeply understand their product margins and their supply chain costs; that's why segmentation becomes so essential to their survival. They are going to have to align supply chain capabilities to prioritized customer groups to stay in business and earn a profit. "With the costs of fulfilling on-line sales for home delivery or 'click and collect' being higher than store replenishment costs this will lead to an erosion of retail net margins and profitability unless retailers take action," said Higgins of LCP Consulting.

The competition in retailing is only going to get fiercer in the decade ahead as consumers use personal technology to shop online for the lowest possible price. And if the item is not available, the consumers won't wait for the company to restock the product. They'll simply go online and find another company that has the item.

To make it in this new environment, retailers are going to have to provide "frictionless selling," according to Michael Zakkour, a principal with Tomkins International Consulting. The idea behind frictionless selling is to make it as easy and as convenient as possible for the consumer to buy, return, and engage with a product brand or retailer. To do that, Zakkour said that retailers and product makers will have to remove barriers, intermediation, and complication from the buying process.

In Zakkour's view, a brand that offers a "true" omnichannel would enable frictionless selling. The customer could buy a company's products in its owned and operated retail stores, at other retailer's outlets, and online and overseas. The brand could provide customers with 2-day, next-day, or same-day delivery. It would offer free delivery and free returns of unwanted merchandise. "Frictionless selling is about creating a multipronged, multi-tech, two-way street for selling, buying and returning," he told me.

Along with making purchases hassle-free, omnichannel retailers will still have to provide a shopping experience to generate excitement among buyers. For web merchants, providing the shopping experience remains a challenge. Although at some point it may be possible to virtually touch and smell a product, that's not feasible right now. But physical contact is just one aspect of the shopping experience, which includes atmosphere, interaction with helpful knowledgeable store clerks, and even a sense of excitement that comes from taking part in an event. To include the shopping experience in their frictionless selling approach, retailers may have to revamp their retail store layouts, said Zakkour on a conference call hosted by Stifel Nicolaus. A store building could be subdivided into special areas. One section would be used for order fulfillment, another for retail transactions, and another as a "brand experience hub" for shoppers to interact and engage with products.

The problem facing retailers is that they have to adapt to dramatic changes in shopping behaviors while at the same time figuring how to manage costs to maintain margins. And that's why retailers will have to set up "protean" supply chains to survive in a new landscape for retailing where a demanding consumer becomes an impatient king. Supply chain executives engaged in e-commerce will have to become adept in making frequent, periodic adjustments. They will have to use supply chain design software to model the impact of changes such as shutting down a retail store and converting it into a fulfillment center to deliver online-ordered merchandise in a key metro market. They will have to use network model software to examine the viability of converting a retail outlet into a showroom store, where customers can touch and see the product but have the good delivered to their house rather than taking it home. They'll have to weigh the costs of providing same-day fulfillment and same-day delivery against the yields from an online sale. In some cases, that might mean same-day delivery, and in other cases, it might mean giving a customer the option of same-day delivery if he or she is willing to pick up the transportation expense.

The ultimate solution to meeting these heightened customer expectation will require a much broader approach than just expediting picking in the distribution or fulfillment center. It will require a much broader approach than expediting delivery to the customer. Ensuring supply to satisfy consumption can't rest on the shoulders of the company with the customer interface. It will have to involve all parties in the supply chain working together. It will require the retailers and all of their suppliers working together as a team. That's why the retailer will have to deploy a control tower as a way to coordinate all parties in the supply chain. Control towers first emerged as a way to control risk in a global supply chain, but their ability to facilitate rapid readjustment makes them an ideal solution for omnichannel and e-commerce management. Because control towers are such an important element of the protean supply chain, the topic will be discussed in detail in the next chapter.

BIBLIOGRAPHY

Amazon to hire 5000 full-times to fill posts at 17 fulfillment centers. *DC Velocity*. Published online July 29, 2013.

Amazon and the United States Postal Service now deliver packages on Sunday, making every day an Amazon Prime delivery day. Amazon press release; November 13, 2013.

Shelly Banjo. Wal-Mart's e stumble. *Wall Street Journal*; June 10, 2013.

Steve Banker and James Cooke. Stores: The weak link in omnichannel distribution. *DC Velocity*. Published online August 5, 2013.

B-roll: Walmart to Go. Online media video for journalists. Wal-Mart Stores Inc. Published online October 15, 2013.

Stephanie Clifford and Claire Cain Miller. Instantly yours, for a fee. *New York Times*; December 27, 2012.

James A. Cooke. Software that eases DC labor pains. *DC Velocity*. Published online November 14, 2011.

James A. Cooke. New kids on the LMS block. *DC Velocity*. Published online February 21, 2012.

James A. Cooke. Is your DC struggling with fulfillment? Consider DOM software. *DC Velocity*. Published online May 14, 2012.

James A. Cooke. Robots: Coming soon to a DC near you (really!). *DC Velocity*. Published online May 13, 2013.

James A. Cooke. How Chico's became "channel-agnostic." *DC Velocity*. Published online August 5, 2013.

James A. Cooke. Omnichannel requires a robotic solution. *DC Velocity*. Published online August 5, 2013.

James A. Cooke. Retailer stores can't handle omnichannel fulfillment on their own. *Supply Chain Quarterly*. Published online August 28, 2013.

Global spending on urban logistics will more than double in a decade. *Supply Chain Quarterly*; Quarter 2, 2013.

Stuart Higgins. Omnichannel fulfillment . . . key strategies and tactics. LCP Consulting presentation; October 2013.

Lauren Johnson. US ecommerce sales to hit $370B by 2017, pushed by mobile: Forrester. *Mobile Commerce Daily*; March 18, 2013.

Chris Jones. Home delivery success starts at the order. *DC Velocity* blog. Posted online January 16, 2013.

Chris Jones. Amazon home delivery: It's about choice, not same day delivery. *DC Velocity* blog. Posted online February 6, 2013.

Chris Jones. The last mile is the last word. *DC Velocity* blog. Posted online June 14, 2013.

Chris Jones. Don't wait for the inevitable in omni-channel retailing. *DC Velocity* blog. Posted online October 21, 2013.

Matt Kulp. Material handling equipment for multichannel success. *Supply Chain Quarterly*; Quarter 4, 2013.

Helgi Thor Legi. Same-day fulfillment: It's a balancing act. *DC Velocity* webcast; October 29, 2013.

Sarah Perez. Walmart expands same-day grocery delivery to Denver. *TechCrunch*. Published online October 15, 2013.

Al Sambar. Beyond omni-channel: Meeting future customer expectations. Presentation for the Retail Supply Chain Logistics Conference; 2013.

John Seidl. Kurt Salmon ship from store point of view. Kurt Salmon Inc. White paper, April 4, 2013.

CHAPTER 7

CONTROL TOWERS

Global supply chains are very vulnerable to breaks. When an undersea earthquake off the coast of Japan triggered a tsunami on March 11, 2011, the surging waters damaged shoreline communities and set off a partial meltdown of a nuclear reactor. As a result of that disaster, many factories in that country were shut down, impacting companies in other parts of the world. The *New York Times* (March 19, 2011) reported that a General Motors plants in Louisiana had to shut down temporally for lack of parts from Japanese suppliers. Another automobile maker in the United States, Ford Motor Co., reportedly also had problems getting red and black pigments for its vehicles from Japanese suppliers impacted by that disaster. In fact, a 2011 survey of 559 companies in 62 countries by the Zurich Financial Services Group and the U.K. Business Continuity Institute found that 20% of survey respondents were affected by the March 2011 earthquake and tsunami in Japan.

In the same year of 2011, summer monsoon rains produced huge fall floods in Thailand that also had a negative impact on global supply chains. The *New York Times* (November 6, 2011) reported that more than 1000 factories in Thailand were swamped, impacting

Protean Supply Chains: Ten Dynamics of Supply and Demand Alignment, First Edition.
James A. Cooke.
© 2014 John Wiley & Sons, Inc. Published 2014 by John Wiley & Sons, Inc.

the production of computer components and affecting supply chains for computer makers.

Weather-related events play havoc with today's far-flung supply chains that span the globe. The same Zurich Financial and Continuity Institute survey cited earlier found that 51% of respondents in the study suffered a weather-related supply chain break in 2011.

But extreme weather or geophysical events like earthquakes aren't the only causes that give rise to a supply chain break. Companies making or sourcing products at long distances from the point of sale or consumption face the possibility of disruptions from political upheavals, geopolitical conflicts, power shortages, pandemics, terrorism, and labor strikes. Even a sudden turnover in employees can have a costly negative impact on the operation of a global supply chain.

In their 2011 survey, the Zurich Financial Services Group and Business Continuity Institute reported that 32% of respondents had had a supply chain disruption that cost their enterprise revenue. Moreover, 17% of those surveyed said that the financial costs from their single and largest single supply chain disruption alone amounted to $1 million euros or more.

Other studies have also noted the dire financial consequences from supply chain breaks. A 2009 study sponsored by PriceWaterhouseCoopers and done by Vinod Singhal of the Georgia Institute of Technology established a solid correlation between supply chain disruptions and negative financial performance. Singhal examined the finances of 600 U.S. companies over the period of a decade. He compared the financial performance of companies that had suffered a supply chain break against a group with a stable supply chain. The bases for his analysis on disruptive impacts were announcements reported in the *Wall Street Journal* or the Dow Jones News Service between 1998 and 2007.

His findings were rather startling. Singhal discovered that the average share price of companies hit by a supply chain disruption fell 9% during the 2-day period encompassing the day before and the day of the announcement. The impact from the blow wasn't just felt for a couple of days. The negative effects of those announcements lingered for quite some time afterward. As matter of fact, 1 year after a supply chain break, company profits were still being impacted. Singhal reported that companies who had experienced disruptions saw their median return on sales ratio—or operating profit margin—drop 4 percentage points, going from 10.5% to 6.4%.

The negative impact from supply chain disruptions on stock performance isn't unique to America. In the article "How to Recognize and Reduce Risk" (*Supply Chain Quarterly*; Quarter 3, 2013), UTI executive Mudasar Mohamed wrote that the Japanese earthquake and tsunami resulted in the Nikkei Index dropping by more than 17% in the 3 days following those disasters.

And the toll from supply chain breaks keeps going up. A survey conducted in 2012 and released in 2013 by Deloitte Consulting LLP stated that some 53% of the 600 supply chain executives canvassed across all industries said supply chain breaks had become more expensive in the past 3 years. That study also found that breaks were especially costly for supply chain operations in such sectors as technology, industrial products, and diversified manufacturing. In the release on the study, Kelly Marchese, a principal at Deloitte Consulting, aptly observed: "Supply chains are increasingly complex and their interlinked, global nature makes them more vulnerable to a range of risk."

For companies with lean supply chains or products with short shelf life, disruptions can be financially onerous because they can impact the flow of parts or products in a global supply chain. Disruptions can result in such outcomes as product quality failure and regulatory noncompliance. If a company needs a certain part, and there's a shortage of them, prices could rise dramatically, resulting in product margin erosion. Product margins can also be eroded if companies are forced to sell goods quickly due to a shortened life cycle or mark down products due to their inferior quality. Worst of all, a supply chain disruption can tarnish a company's reputation. And in a world where brand allows companies to obtain premium prices for its goods, the loss of reputation can hang on for years past the date of the supply chain break.

Speaking at the 2011 Council of Supply Chain Management Professionals (CSCMP) Conference in Europe, Dan Mahoney, a supply chain executive with Intel International B.V., told the gathering: "When a supply chain breaks, you get lower service levels, higher costs and inventory or longer lead times. They will be noticed by customers or shareholders."

Certainly, no supply chain executive wants the notoriety that comes from a supply chain break on his or her watch. Perhaps due to the 24-hour, around-the-clock news coverage on television and on the Internet, there's a feeling that calamitous events seem to be happening with greater frequency. One year it's Hurricane Sandy in the news spotlight; the next year it's a typhoon in the Philippines. There's no pattern whatsoever as to the location and timing of those tragedies, reinforcing the sense of randomness.

If the world today seems a place where chaos is lurking and mayhem is waiting to strike, companies need to assume the worst. Luck will eventually run out. Murphy will be proven right, and anything that can go wrong will go wrong. Given the mood of the current times, it's no wonder that so many supply chain executives have a sense of dread about what could be coming their way. As Timothy Carroll, IBM's vice president of global execution for its integrated supply chain, said in the article "From Many, One: IBM's Unified Supply Chain" (*Supply Chain Quarterly*; Quarter 4, 2012): "Most supply chain chiefs don't worry about what they know. They worry about what they don't know."

Given that the odds seemed stacked toward an eventual disaster, a company with manufacturing and distribution spanning continents will sooner rather than later see its supply chain suffer a break. To survive those breakdowns, a supply chain has to be resilient. It has to have the ability to quickly recover from an interruption.

That's why many companies prepare for the eventuality of disaster. They accept its inevitability and develop a plan for how the supply chain will respond. They envision disruption scenarios and devise contingency plans for how to respond if the supply chain breaks. Generally speaking, companies without any sort of plan or risk mitigation program put themselves at greater risk than those who have one. "This increased complexity [of supply chains], coupled with greater frequency of disruptive events such as geopolitical events and natural disasters, presents a precarious situation for companies without a solid risk program in place," said Marchese of Deloitte Consulting.

Usually, these plans identify backup suppliers and backup carriers. In the case of strategic suppliers whose products can't be readily acquired elsewhere, a company can insist that a key provider of parts or materials even have a contingency plan of its own. The supplier's plan should spell out how that supplier itself can procure additional capacity during a disruption.

Along with scenario planning, companies should get a handle on the degree of risk facing them by engaging in segmentation and risk quantification. Mohamed in the aforementioned article advised companies to segment their supply chain, breaking it down by product lines and distribution channels. Then a company could assign a priority to each supply chain segment based on such factors as revenue, gross margin, stock keeping units (SKUs), unit volume, and, lastly, strategic importance. After the company has segmented its supply chain, it can also quantify the risk of an occurrence. The results of this analysis can be then used to develop contingency plans to address a variety of scenarios for defined segments.

Although it's certainly prudent for a business to develop a contingency plan for possible supply chain disruptions, that document can only tell a company what to do after an event has fractured its supply chain. What companies in a volatile world have to be able to do is take corrective action at the moment the supply chain break occurs. They need the capability for a dynamic response. And that's why companies need to adopt a control tower strategy as a way to deal with supply chain risk.

In the field of aviation, control towers use radar to monitor the coming and going of aircraft to and from an airport. The air traffic controllers in the tower monitoring the skies can speed up or slow down aircraft en route to the airfield. They can even reroute planes when a storm shuts down an airport. The concept behind a control tower for a supply chain is similar in that it monitors orders, inventory, shipments, forecasts, and their status throughout the globe. The staff running the control tower can see activity related to inventory, orders, and so on, on their consoles and computer screens. Dashboards are often used to provide visual representations on the computer screen about ongoing activities that are being monitored. Dashboards highlight important indicators such as shipment or inventory status and focus attention on key trends, comparisons, and exceptions. "The adoption of a supply chain control tower enables firms to better understand efficiency levels," said Martin O'Grady, director of Outsource Management Service Ltd. in the United Kingdom. "There is no other way to run a network without true visibility otherwise you have to ask: 'what are you managing?' "

But a supply chain control tower does more than just provide visibility. It can facilitate rapid decision making when a crisis occurs. It allows a company to orchestrate its partners to take collective action. Say, for instance, a truck is on its way to a port with a critical parts delivery when an ocean storm delays the port call of a container ship. Because the factory needs that parts delivery to meet its commitment to a major customer for a product order, the control tower can divert another parts shipment en route and thus ensure steady production. As supply chains have grown more complex due to product proliferation, outsourced manufacturing and contract logistics, and the expansion of sales channels, communication among the various organization in an extended supply chain has become increasingly difficult across multiple time zones and geographies when a problem requires a quick, coordinated resolution. A control tower provides a mechanism for group action.

To date, companies that have set up control towers generally do so as a way to better manage their carrier or supplier base. The party running the control tower and coordinating the group response to avert a problem is often the brand owner of the product or the channel master. Those two usually have the most at stake in terms of reputation when the supply chain fails to function properly. Members of the general public don't know the identities of the suppliers behind a name-brand product; if a screwup by the supplier results in inferior product performance, the buying public blames the problem on the manufacturer whose name adorns the product. Likewise, if a supply chain break prevents timely delivery of a product promoted in the media for a special store sale, the public does not blame the supplier. The public squarely places blame for the failure on the channel master—the retailer or grocer who controls the sale of the product.

To fashion a coordinated response, control towers have to gather information from each of the supply chain partners. The underlying technology is an information platform or hub that the supply chain partners tie into. Connection to the tower provides a way to reduce or even possibly eliminate information latency, the time delays that occur when companies exchange data with one another. Ideally, the tower should ensure that flow of information moves in sync with the flow of products and parts through the supply chain. In the past, "because the partners were not connected, problems were not seen until after they occurred," said Alain Poirier, a vice president of business development at the firm Tompkins International on a webcast titled "Supply Chain Control Towers: Separating Fact from Fiction."

The essential work in setting up a control tower is linking the supply chain partners' information systems together to a central data hub or platform. Because it's rare for all suppliers, carriers, and other parties to use the same software from the same vendor, special software known as middleware is needed to translate messages and information between disparate applications. Middleware is used to connect the tower platform with enterprise resource planning (ERP) systems, warehouse management systems (WMSs), transportation management systems (TMSs), manufacturing execution systems (MESs), global trade management (GTM) systems, and advanced planning and scheduling (APS) systems. The tower hub can be set up with computers on a company's premises, or it can be set up in the "cloud" with access via the Internet. Nowadays, a cloud-based control tower has become the standard because of the low implementation costs and ease of connecting partners with that approach, Capgemini Consultant Rob van Doesburg told me.

The control tower establishes a central data repository for information. Although towers are principally set up for monitoring and management, a side benefit to having a central data repository is that information for reports are stored in one place. That makes it easier to access records that might be required in the event of an audit. Should the channel master or brand owner need to conduct a review of the supplier's performance or the carrier's performance, the records are stored in one place, eliminating the need to track down information.

The mechanism of a control tower allows for central planning of the extended enterprise. Since each supplier has its own plans, and there are generally multiple suppliers in the supply chain, a control tower offers a way to impose discipline on an extended enterprise. That's because the control tower can look across the inventory holdings of all suppliers. In the past, it was not uncommon for a supplier to shield some stock from the manufacturer, holding back parts that might be needed and even increasing lead times for inbound materials. In doing so, the supplier was tending to its interest often to the detriment of the channel master or the brand owner. A control tower thus minimizes suboptimization of the supply chain. "A control tower if coupled with multiparty networks can deliver collaboration planning and execution," said Poirier during a webcast on this topic.

The control tower connects planning software with execution software. That enables a rapid response to unfolding events. By feeding real-time information on events—say, a delay of an inbound shipment to a factory from an impending ocean storm as in the earlier example—the planning software could immediately propose a substitute delivery using a second supplier to maintain continual production. In the past, reassigning inbound parts delivery was difficult to do on the fly as a switch involved production planners making numerous calls and sending numerous e-mails to determine even the feasibility of such a move.

For companies to fashion a collective response to demand signals, they have to have a "single version of the truth" for reasons discussed in the second chapter on demand-driven supply chains. Supply chain partners all have to be on the same page, so to speak, on what the data mean. A control tower allows the channel master or the brand owner to impose discipline on data repository, which can be used by application for both planning and execution. Parties in the extended enterprise have to agree or at least accept the channel master's or brand owner's definition for demand and estimates for future sales. There's no room for debate over the interpretation of the demand signal as that would defeat a quick response.

Because it's collecting data on activities, the control tower allows the channel master or brand owner to monitor and manage the performance of its supply chain partners. The control tower can employ scorecards in this regard. Those scorecards could be used to manage supplier compliance with service standards. For instance, it could ensure that the supplier follows the routing guides, the use of particular carriers for shipments to specified locations. But it's not just suppliers that the tower can monitor with scorecards. Scorecarding can be used to measure whether the carriers are honoring their time commitments for delivery. By monitoring suppliers and carriers, the tower can assist with overall performance improvements, enhancing the overall supply chain flow.

Control towers usually employ event management software that flags deviations from the plan. The container shipment traversing across the Pacific Ocean on the steamship line has not reached its scheduled milestone on the journey. The work-in-process inventory has dipped below a specified level such that there's less than a day of supply on hand. The event management software can send out an alert to the staff working the control tower to prompt them to look into the matter and to take corrective action. In that regard, it supports the concept of management by exception, allowing a company to devote limited resources on addressing activities that deviate from the norm.

To enable a quick but appropriate response control, towers should be equipped with analytical software that applies business intelligence to solving problems. Since the purpose of the tower is to maintain the correct flow, it has to be capable of rapid problem resolution during or shortly after the moment of crisis. Analytics allows the control tower staff to diagnose the problem in a short time and to offer a solution. In an article titled "A Commanding View" (*Supply*

Chain Quarterly; Quarter 3, 2011), Karen Butner said that the use of advanced analytics in control towers would give companies the means to decipher complex streams of data and to make "immediate" decisions based on business rules. "This makes it possible to synchronize end-to-end transactions with intelligence to optimize inventory at all phases, and to predict demand variations," wrote Butner, a director of research at the business think tank, the IBM Institute for Business Value.

Because the tower affords end-to-end visibility, marries planning and execution functions, and uses business intelligence, this mechanism allows the brand owner or channel master to exert more control over the process of order management from start to finish. When a purchase or sales order comes in for a product, the control tower operator can examine its suppliers' inventory to determine parts or component availability. It can reassign inventory if need be and adjust manufacturing schedules as well. As parts are used up in manufacturing, inventory levels at suppliers can be monitored to determine the need for reordering. The tower can then track product delivery from the factory or distribution center (DC) to ensure that customer commitments are met. Most important of all, should something happen, the tower can facilitate changes to the flow.

Companies with control towers generally set them up in one location with a dedicated staff. That team often consists of cross-functional experts since the control tower response can affect manufacturing, procurement, and logistics. By the way, companies can set up a control tower on their own, using special software provided by vendors. Or they can farm out the control tower function to a third-party logistics (3PL) provider. For the most part, 3PLs have been involved in running control towers whose main focus has revolved around routing shipments, either inbound or outbound, from plants or DCs. Some of the first control towers were set up when companies wanted to consolidate their contract logistics operations under one roof. The lead logistics provider managing the other 3PLs set up a control tower to oversee distribution.

What types of companies have deployed control towers so far? According to van Doesburg of Capgemini, manufacturers with many suppliers or outsourced activities have implemented control towers as well as manufacturers with central DCs covering a large area such as Europe who want outbound visibility from the port of entrance to the final customer. Finally, retailers looking for inbound visibility from their suppliers to their DC have also set up towers.

van Doesburg said companies justify the expense of a control tower solution on four grounds: savings on transportation costs, savings on operational costs, improvement quality, and a reduction of working capital.

Companies implementing control towers tend to have complex supply chains, high product volatility, and often low margins with high volatility, said Kirk Monroe, a vice president of product at Kinaxis, which provides control tower software. Monroe added that companies in such industries as high-tech electronics, industrial, and automotive have shown the most interest in this concept.

One company using control towers is the Procter & Gamble (P&G). In her presentation at the 2011 CSCMP Europe Conference, P&G executive Namrata P. Patel described how her company had established control towers with the help of logistics service providers to manage distribution in Europe and Africa. One control tower handles Central and Eastern Europe, while the other serves Africa and the Middle East. Those control towers manage product flow to the distributors that the company uses to sell produce in those regions of the world. Patel further stated that the control towers have paid for themselves by saving the company money on distribution costs.

One of the earliest deployments of a control tower was in 2003 by The Netherlands-based company e-Logistics Control B.V. for Swedish vehicle manufacturer Scania AB. A maker of trucks, buses, and diesel engines, Scania decided to have a single lead logistic provider to manage inbound transportation from the manufacturer's European suppliers to its European production units, its spare parts warehouses, and its hub for overseas transportation. A case study about this is published on the e-Logistics Control website.

The control tower facilitates electronic communication between Scania, its suppliers, and carriers. It oversees the movement of 3000 orders per day transported by more than 100 carriers. The control tower operation ensures that the right mode of transport and equipment is used for each shipment. e-Logistics Control states that it provides visibility down to part number level. As a result of the accuracy of the deployment information and 99% predictability of shipments, there has been costs savings in excess of 10%. e-Logistics Control provides a similar control tower service for several other companies such as Electrolux, IAC, Inalfa, Tenneco, and Amcor Flexibles.

The Scania control tower is an example of one used to manage logistics operations. This mechanism is also to better marshal the flow of inventory. One example of a company doing that is First Solar, a client of software vendor Kinaxis. The Tempe, Arizona-based company makes solar panels as well as builds and manages solar power plants. The company initially set up the tower to move to a more demand-driven supply chain. First Solar used the tower to measure inventory levels and balance inventory to meet targets. As a result, First Solar was able to monitor inventory "positions in minutes, not days," according to a case study available on the Kinaxis website.

That visibility meant that First Solar was able to provide its sales team with exact figures on what inventory was available to sell. Due to the enhanced visibility within 3 months of deployment, First Solar witnessed a reduction in overall inventory. The company was also able to put in triggers in place to indicate how much supply it should produce based on demand. Shellie Molina, a vice president at First Solar, was quoted in a Kinaxis case study as saying, "Our overall inventory reduction was reduced by 8 percent and our aged finished goods inventory was reduced by 84 percent."

After the control tower was set up to address demand planning, it was extended to supply and project management. That extension jived with shift in business strategy. First Solar expanded in 2011 from just selling individual solar panels to building utility-scale power plants. Since construction projects are subject to weather delays and permitting issues, the control tower gives First Solar visibility into how a supply chain disruption could impact a project. First Solar is also using simulation capability to model different potential scenarios to understand the implications of impediment such as longer permitting.

Agilent Technologies Inc. is another company that put a control tower up for its Electronic Manufacturing Group (EMG) in 2011. The company, which makes instruments of various kinds, has more than 1100 suppliers, most of which are based in Asia. The control tower connects five of its plants, five contract manufacturers, and majority of suppliers to get a big picture view of its parts inventory. Michael Tan, Agilent's supply chain operations director in EMG based in Penang, Malaysia, said the tower had visibility over 92% of the components used in the building EMG products.

Agilent uses the control tower in its day-to-day operations to determine whether it has the components on hand to meet a specified time commitment for customer delivery. The tower, which uses Kinaxis software, runs a simulation to quickly determine whether it has the parts or can get them from other sources to meet a customer's timetable. In my article, "The Power of a Control Tower" (*Supply Chain Quarterly*; Quarter 4, 2013), Tan said, "The tower lets you know how much you have on hand and how fast you can those parts into the factory that produces the product for the final customer."

Aside from bid assessments, Agilent has used the tower for crisis management situations. For example, in 2011, Agilent was able to deal with problems resulting from the Thailand flooding, even though it had a contract manufacturer and suppliers in that country. The control tower enabled Agilent to swing into action during the Thailand crisis and to find alternative parts. The tower used software to simulate material shortages. Armed with that information, Agilent was able to work on finding alternative sources of supply or, in some cases, to redesign products to use parts that were on hand. Because of its control tower, Agilent was able to respond to loss of suppliers and to meet most of its customer commitments. "The control tower helps us to be able to capture all components during a shortage so we can come up with risk mitigation action," Tan said.

Some companies are even using their control towers to get a glimpse into unfolding events and then take action before supply chain problems occur. One of those is Dell Inc. The computer maker operates a control tower—or what it calls a "command center"—for its parts and field service delivery. What's so fascinating about Dell—what makes its operation so cutting-edge—is that the company is using the command center to take action in the face of an impending event.

Dell gets a fair amount of its revenue from after-sales support of its products. To ensure topnotch customer support, the company set up command centers. In 2012, when I wrote my article "Inside Dell's Global Command Centers" (*Supply Chain Quarterly*; Quarter 3, 2012), Dell had centers in Austin, Texas; Limerick, Ireland; Kawasaki, Japan; Xiamen, China; and Penang, Malaysia. On the walls of the command centers are an array of mounted screens showing maps, weather, news, and service requests from customers. Dell uses its own proprietary technology platform to provide real-time information updates for those monitors.

Staff at these five command centers route spare parts from more than 600 depots around the world to customers. They'll also dispatch technicians to a customer's site. But what's unique about the command center is the use of special software for its visibility platform. The software takes data from Dell's internal computer systems as well as that of its partners and then feeds it into a rules engine. The engine can detect exceptions and then alert a command center staffer about the need to take a corrective action.

Those special applications also connect to a geographic data system. As a result, Dell can take into account current weather information to assess the potential impact of a storm on its parts shipments. Steve Sturr, executive director of global services at Dell, told me that it's possible for Dell to warn a customer of an impending storm and even to recommend preparatory actions. Not only can Dell respond to supply chain disruptions quickly but the company can also reach out to customers and offer assistance.

Dell is clearly a leader when it comes to control towers. For companies operating global chains, or even a regional supply theater, control towers will be critical to enabling a proactive response to a supply chain disruption. The towers provide an end-to-end view of what's happening in the supply chain. Without real-time visibility, it's impossible to respond to a crisis except after the fact.

Having the control tower incorporate analytics into its operation—as Dell did—gives a company the ability to actually manage risk in its supply chain even though a control tower gives the ability to align supply with demand across the entire supply chain; this technology is still leading edge as only a handful of big multinational companies have deployed control towers to oversee their far-flung supply chains. But the idea is starting to attract wider interest. In the 2012 Annual Supply Chain Study by Capgemini Consulting, 57% of 350 companies taking part in the research said that visibility improvement such as control towers were high on the list of supply chain projects to undertake.

Not only are commercial enterprises interested in deploying control towers but federal agencies are also looking at this concept to oversee supplies. "Commercial organizations and federal agencies are becoming increasingly focused on obtaining the ability to be proactive to changes in their businesses and responding to minimize the impact to customer service or cost," Ron Ash, managing director of Accenture's Federal Services told me. Ash noted that an organization can set up a control tower with a broad focus to encompass the entire supply chain or narrow the focus to a specific activity such as sourcing or transportation planning. "Most organizations start with a proof of concept to validate the idea and value, then move throughout the supply chain for a broader control tower."

For supply chain chiefs charged with risk management, control towers allow a company to go from reacting to anticipating. If it's equipped with business intelligence, a control tower allows a company to pick up on early trends that could lead to disaster or a supply chain break. Dell's experience shows how companies can become proactive in handling problems.

For suppliers and carriers, connecting to several control towers raises some issues as they are required to devote resources to multiple connections and share proprietary information—something that the business may be reluctant or unwilling to do. Since control tower visibility is dependent on access to data, the company setting such a center up for a channel may want to make information sharing a requirement for suppliers and even customers a requirement for doing business with it.

Speaking at the Eyefortransport 2013 Conference on Big Data in the Supply Chain in Chicago, Douglas Kent, vice president of global supply chain at distributor Avnet Inc., said control towers could provide manufacturers with "multitier visibility" of suppliers. Kent went on to say that companies want suppliers to merely provide them with information already in their possession. "We're not asking them [suppliers] to create information," he said. "We just want to use the data on our doorstep and bring it in house."

At the moment, companies are using control towers as a way to manage risk in logistics and supply. But the purpose behind control towers will expand to other areas of the supply chain in the future. "Risk mitigation was the first area of focus for most customers," Ash told me, "but shortly after that they began coming up with new and innovative ways to help drive sales growth, increase customer satisfaction, reduce costs and increase flexibility and reactiveness of their supply chains."

One way for companies to extend the control tower concept would be to set up centers at their distribution facilities to better manage warehouse workflows. What happens outside the four walls of a warehouse can impact the operations inside the building. If a truck runs into traffic on a highway, it can miss its delivery window, and that has implications for scheduling and assigning the warehouse labor force.

At the moment, DCs don't have the visibility to adjust their plans to changes in events.

That's because a DC uses warehouse management software to oversee warehouse operations, transportation management to oversee shipping to and from the warehouse, and even yard management software to oversee tractor and trailer movements in the warehouse yard. In a dynamic environment, a warehouse in a protean supply chain has to be able to react and respond to changes in events.

A control tower at the DC level would take information from various applications such as WMS, TMS, and an order management system to form a big picture of operations inside and outside the warehouse walls. Just like a radar screen that changes with the flow of aircraft, the DC picture would change to reflect events in real time.

Up-to-the-minute visibility means that a company can make adjustments to their DC operations. A control tower would facilitate the reassignment of warehouse workers and equipment to minimize disruption. If a truck will not be able to arrive on schedule because a freeway accident clogged the highways, then the control tower could use the notification delay to change the plan of daily activities. In that example, workers could be reassigned to load a truck while waiting for the delayed delivery. The point here is that the control tower at DC allows for a rapid response to keep the warehouse operation "humming." Although only one large consumer packaged goods company—to my knowledge—is currently running a control tower for its DC, other companies will need to consider this type of deployment. A control tower in DC makes sense for a company that runs a protean supply chain.

Improving DC productivity is just one area of supply chains where control towers could be applied. Control towers may actually offer the ultimate solution for making an omnichannel strategy work properly. That mechanism would permit retailers and consumer packaged goods companies to coordinate their responses to demand variation up and down the supply chain to online and physical stores. In a demand-driven supply chain, retailers will need visibility as to where items are selling and where replenishment shipments are going. Control towers would give retailers the way to divert in real-time shipments—inventory in motion—and to reposition products in stores or DCs as demand patterns shift.

But replenishment modification is only one facet to running a demand-driven supply chain for omnichannel retailing. Retailers must be able to get manufacturers to change course. If a retailer picks up a demand signal that signifies a new buying trend, it can revise its forecast as to what to make or stock. But a change in plan must be lead to action. A plan change must be communicated rapidly to enable a sudden shift in execution. If demand picks up for particular item, the retailer must notify the manufacturer to increase production. Conversely, if demand slows down for a product, the retailer must convey that as well so the manufacturer can adjust production. Control towers give a mechanism for that to happen, and happen quickly.

Control towers also offer retailers a way to manage the competing and conflicting demands that come from servicing online and physical stores. Say the retailer faces a surge in sales for a particular product. That leaves the retailer facing a "Sophie's choice" as to whether to put remaining inventoried products in an online fulfillment center or on the store shelves to catch the eyes of shoppers. A control tower would enable the retailer to scan the current production and inventory of all of its suppliers. A control tower might determine that a tier-one supplier to the product manufacturer has available components but shifting components would jeopardize production on another item. In addition, the control tower might allow the retailer to determine whether a second supplier has the parts or materials on hand to produce a substitute product to satisfy some of the additional demand. Once the decision has been made on how to arrange supply, hopefully on the basis of a segmentation strategy and on the basis of analytics, then the control tower will be able to coordinate all the relevant suppliers to work together to make and ship the product that's most needed.

To my knowledge, no retailer has done this to date, although software vendors have told me that a couple have considered this approach. Although faster throughput enabled by technology in fulfillment centers and the application of DC practices to in-store order picking will help retailers handle online orders more swiftly, they don't provide a supply chain solution. They don't enable a responsive connection between point-of-sale demand and production. A control tower is the only way to provide this.

To meet variable channel demand in real time thus requires end-to-end visibility of inventory and response capability. Although retailers will need the ability to see into their suppliers' inventory holdings, they will need the ability to tell a factory to change production at a moment's notice. Indeed, success for an omnichannel retailer or web merchant could very well depend on its working with its suppliers to take coordination actions to consumer demand shifts. Control towers will give the omnichannel retailers the nimbleness and supply chain mutability they need to respond with the products that fickle consumers may be demanding at a particular moment in time. "Linking control towers with network orchestration synchronizes all trading partners so that they are all focused on near-real-time sales, operating the supply chain in concert," said Poirier on the aforementioned webcast.

Mutability, however, would require a different supply chain mindset than what now exists. At present, most manufacturers have a make-to-stock mindset and they promise retailers an allotment from their available inventory. Manufacturers would have to begin promising and reserving the "capability" to make products upon request. The actual product to be made, however, would be determined at a specified time window to meet the retailer's variable demand. If true demand begins driving replenishment and production, the item made days earlier and sitting in warehouse inventory may not be what's needed to satisfy consumer demand. Mutability necessitates a dramatic change in manufacturing practices. Manufacturers would no longer make products and set them aside in a warehouse—a practice called available to promise. Instead, the manufacturer would engage in the practice of "capable to promise" or setting aside a production slot in the factory.

If manufacturers do switch to capable to promise production, Trevor Miles, a vice president of thought leadership at Kinaxis, said that for white goods and electronics, more manufacturers would have to engage in a postponement strategy, finalizing the actual makeup of the product at the moment of the order. (By the way, this is exactly what Dell Computer did in its early days, building computers to meet the exact customer specifications.) "If OEMs [original equipment manufacturers] need to supply a bundle or kit as opposed to a FG [finished good], then they to be able to match the supply for each items in the bundle," he explained. "The omnichannel aspect has little impact on this other than that often the demand is more variable and the product portfolio bigger with more variants, the consequence being that carrying finished goods is really risky so many companies should be deploying postponement strategies, which then goes towards capable to promise."

Capable to promise could well become the differentiating attribute for manufacturers in a dynamic business environment rather than the ability to produce goods at the lowest costs. Manufacturers have to become more adept at changing overproduction lines in a world of demand-driven supply chains. At the moment, that poses a huge challenge for product makers. A November 2013 study by Accenture Consulting found global manufacturers lacked the flexibility required for growth. "Manufacturers are finding that their existing operational models are not effective in this new volatile marketplace," said Russ Rasmus, global managing director of the Accenture Manufacturing Practice, "but there is clear evidence that they are responding by making a range of investments to create more flexible and consistent operations in the future." Advances in manufacturing technology such as 3-D printing and intelligent robots, discussed in more detail in a subsequent chapter, may help obtain manufacturers to achieve more agility.

As supply chain production and replenishment become demand-driven, control towers offer the mechanism for retailers, manufacturers, and suppliers to work in concert to match true demand. Unfortunately, today, most companies in global economy lack end-to-end visibility and real-time decision-making capabilities to execute a rapid, flexible response to changing circumstances of supply and demand. A 2013 study by the Hackett Group Inc. looking at more than 100 companies found that "less than half have near-time visibility into customer information and business volumes, and even fewer have the same level of visibility into supplier spend, working capital, financial performance and forecasts, and risk."

That will have to change. Because supply chains in a dynamic world need visibility and responsiveness, control towers become an essential component in operating a protean supply chain. To date, most companies have used control towers or command centers to obtain more flexibility in managing their suppliers or carriers. They deploy control towers as a risk mitigation strategy, justifying the investment based on the high cost of a supply chain disruption. In the future, companies will have to justify the investment in a control tower based on sales revenue. That's because the tower allows a group of supply chain partners to respond to ongoing events or marketplace happenings in minutes rather than days or hours. In a 24-hour global marketplace ruled by e-commerce, companies no longer have the time to delay in keeping customers happy.

BIBLIOGRAPHY

Gaurav Bhosie, Prashant Kumat, Belinda Griffin-Cryan, Rob van Doesburg, MarieAnne Sparks, and Adrian Paton. Supply chain control tower. Capgemini Consulting white paper. 2011.

Business case Scania; e-Logistics Control case study. E-Logistics Control B.V. Published online 2013.

Karen Butner. A commanding view. *Supply Chain Quarterly*; Quarter 3, 2011.

James A. Cooke. Inside Dell's global command centers. *Supply Chain Quarterly*; Quarter 3, 2012.

James A. Cooke. From many, one: IBM's unified supply chain. *Supply Chain Quarterly*; Quarter 4, 2012.

James A. Cooke. Does your DC need wider vision? *DC Velocity*. Published online August 12, 2013.

James A. Cooke. The power of a control tower. *Supply Chain Quarterly*; Quarter 4, 2013.

Thomas Fuller. Thailand flooding cripples hard-drive suppliers. *New York Times*; November 6, 2011.

Global manufacturers lack flexibility described as critical to their growth, finds Accenture study. News release. Accenture; November 21, 2013.

The high price of supply chain disruptions. *Supply Chain Quarterly*; Quarter 1, 2009.

Kinaxis. How First Solar is achieving end-to-end supply chain visibility. Kinaxis case study. 2013.

Steve Lohr. Stress test for global supply chain. *New York Times*; March 19, 2011.

Kelly Marchese and Siva Paramasivam. The ripple effect how manufacturing and retail executives view the growing challenge of supply chain risk. Deloitte white paper. 2013.

Mudasar Mohamed. How to recognize and reduce risk. *Supply Chain Quarterly*; Quarter 3, 2013.

P&G establishes logistics "control towers." *Supply Chain Quarterly*; Quarter 3, 2011.

Alain Poirier and Mark Zelenak. Control towers: Separating fact from fiction. One Network Enterprises video. 2013.

Fred Sandsmark. Platform for efficiency. *Profit Magazine*; August 2012.

Rob van Doesburg and Ramon Veldhuijzen. Supply chain visibility insight in software solutions. Capgemini Consulting. 2012.

Viewlocity. Supply chain control tower visibility. White paper. Viewlocity; 2013.

Weather is the leading culprit for supply chain disruptions. *Supply Chain Quarterly*; Quarter 4, 2011.

CHAPTER 8

FAST DATA

To run a protean supply chain, companies will have to connect the dots quickly, aligning demand swings with production and distribution. They will have to process demand signals, and then adjust their supply chain operations in response to nascent demands to ensure a steady flow of material and products from the supplier to the end user. To do that, companies will have to make data-driven decisions even faster than they do now using more types of information than they do now.

To make smart on-the-spot adjustments, companies running protean supply chains will have to take advantage of the latest software tools to analyze data in their supply chain operations in real time. Sure, existing analytical or business intelligence software can monitor and measure supply chain activities from production to distribution, from inventory to transportation. But analytics will have to do more than just measure and monitor for efficiency and accountability. Analytical software will have to be able to make connections. It will have to have the ability to spot problems in the making or to detect changes that can cause future problems. And the software will have to pull that information from various repositories of data throughout the supply chain.

Protean Supply Chains: Ten Dynamics of Supply and Demand Alignment, First Edition.
James A. Cooke.
© 2014 John Wiley & Sons, Inc. Published 2014 by John Wiley & Sons, Inc.

Companies have been using analytics to appraise data and to get a better understanding of their business operations since after World War II. And logistics and supply chains have been one of the prime focus areas for the application of analytics. "Applying analytics in supply chain management is not a new idea," wrote Thomas H. Davenport and Jerry O'Dwyer in an article titled "Tap into the Power of Analytics" (*Supply Chain Quarterly*; Quarter 4, 2001). "The U.S. military adopted a variety of logistical models in World War II, and companies adopted related approaches in the postwar period. UPS, for example, established a logistical analytics group in 1954. Since then, many companies have successfully employed analytical approaches to distribution networks, inventory optimization, forecasting, demand planning, risk management, and other applications. Large retailers, such as Wal-Mart Stores and Target, have had considerable success with supply chain analytics, often working in collaboration with suppliers. And carriers like UPS, FedEx, and Schneider National wouldn't dream of managing their operations without a variety of analytical models."

Although it's possible for managers to use a spreadsheet to do a simple analysis comparing a single unit of measurement—say, on-time delivery or units produced per line—it becomes a difficult, time-consuming tedious exercise to gather all the necessary information. It's even more time-consuming to pore over the collection of numbers in spreadsheet cells and to make multiple comparisons to find that "aha" insight. Fortunately, today, many applications used in running the business now provide analytics. Enterprise resource planning systems—the information system backbone of many companies as it carries out financial and business planning using a common database—provide analytics today. Many supply chain execution applications, such as transportation management systems (TMSs) and warehouse management systems, provide analytics as well. There's also special software just written to do an analysis for a specific area.

Most of the analytics in use today in the supply chain describe how operations were done in the past. Indeed, so-called descriptive analytics are used in a number of areas in the supply chain. "What people call *descriptive* is ex post facto analysis of what already happened," said Gartner analyst C. Dwight Klappich.

One current use for descriptive analytics is performance evaluation. Software can be used to compare the performance of contract manufacturers, suppliers, transportation carriers, and third-party logistics companies. For example, a company could set up a scorecard with various measurements or metrics to analyze the performance of third-party logistics (3PL) companies handling their warehousing and transportation. The 3PL could be measured on percentage of orders received on time, the percentage of items damaged, invoice accuracy, and order fulfillment lead time. Analytical software could rate all contract logistics providers and identify which providers are not performing up to par.

Besides evaluating performance, analytics can be used to study costs. To get a better understanding of expenditures in the area of procurement, many companies use spend analysis or spend management software. Companies apply spend analysis to scrutinize suppliers, the cost of supplies (bills of material), and even services. Along with collecting and classifying data on expenditures, this type of software lets a company drill down beneath the surface to identify underlying factors. For example, in the case of rising transportation expenses, the factor pushing up costs could be a hike in fuel surcharges, which are tied into diesel fuel prices and, ultimately, the price of a barrel of oil on the world market.

Although historically used in procurement for costing out materials and parts, spend analysis software has started to be applied to services such as warehousing and transportation to find areas for cost savings. Spend analysis software has also been used to create a template for evaluating carrier bids for transportation lane movements on the basis of price and service criteria. This has proven valuable to retailers who can use spend analysis to evaluate whether they should assume control over inbound transportation, the delivery from the manufacturer's plant to the retailer's distribution center. Manufacturers charge the buyer for freight delivery, sometimes as a separate charge, sometimes rolled into the product's price. And in some situations, when delivery charges are tucked into the product price, the manufacturer even takes a markup on transportation. For the retailer then, in some cases, it can be cheaper to arrange for its own transportation and to have its own carrier pick up the shipment directly from the manufacturer. The results of a spend analysis on inbound transportation allows a retailer to make a fact-based decision on whether it makes economic sense to take over delivery.

Analysis of inbound shipping can also be done by a TMS, a type of execution software used by shippers to manage freight movements. The TMS is especially useful if a shipper wants to compare pricing for goods movements on individual shipping lanes. More advanced versions of TMS can routinely record operational data on each carrier and then automatically apply that data to performance metrics such as on-time delivery or damage in transit.

Because the TMS can look across the entire carrier base, it can pinpoint where a particular carrier may be overcharging on a lane or has a low percentage rate of on-time deliveries. Because many TMS applications already come with built-in analytics, Klappich said the software has ready-made key performance indicators to set up a carrier performance scorecard. That feature can help shippers make smarter decisions in selecting carriers. So if a carrier with low rates had a lousy on-time delivery record, the TMS could be set up to automatically select a different carrier for time-sensitive shipments.

Although a TMS can perform many types of analysis, there's also available special software written just to do detailed freight spend analysis. In the article "How Invacare Slashed Its Freight Spend" in the October 2012 issue of *DC Velocity*, I described how the home medical equipment division of Invacare used special software to do an analysis of its shipping lanes to gain leverage in negotiations with its trucking companies. Truckers these days vary their route charges. They usually charge higher rates on lanes where they don't want shipper business, perhaps because backhaul loads are hard to come by. By the same token, truckers charge lower rates on lanes where they want loads from shippers.

Invacare had to take its freight shipping requirements and check them against the tariff bases of multiple trucking companies. That would have been a nearly impossible task if done manually. Invacare used software from a vendor called RateLinx that analyzes tariffs by lane. By determining those lanes on which a trucker sought business, or didn't, Invacare was in a stronger position to negotiate better terms from its carrier base.

But it's not just transportation where analytics are helping supply chain managers find savings or efficiencies. Many versions of warehouse management software (WMS)—another type of supply chain execution software—have the capability to do descriptive analytics. "Virtually all WMS and TMS have this capability to one degree or another," Klappich said. "Some are better than others and there is more or less the ability to drill up and down in the data. And some are better at presenting the information but they all have it."

Aside from analytics in WMS packages, there are other business intelligence tools for examining warehouse operations. Labor management software (discussed in Chapter 6) often has analytics to appraise worker performance in doing such tasks as picking, receiving, and putaway. One other interesting piece of analytical software I've written about examines throughput in an automated distribution center. This particular tool can be used to determine why one warehouse has a lower level of throughput than other facilities in the company supply chain with the same equipment. This particular software can analyze problems that occur with "side-by-sides" in an automated sortation system. Side-by-sides occur when a smaller package gets squeezed up against another larger package, resulting in both products being mistaken as a single unit and diverted together down a lane.

Keep in mind that when thousands of packages flow along a high-speed conveyor line, the bar code has to be placed on the right location on the box for a laser, mounted overhead, to read the symbol quickly and have the computer determine where to divert the box. Pictures of the bar code symbols can be taken and those photo images stored inside a computer database. An analysis of those images can identify whether the suppliers are shipping boxes that are in "bar code" compliance, meaning that the symbols are applied in a manner on the box as requested by the receiver. The analysis can determine that if bar codes on the cartons were clear and located properly, the gap between the boxes on the conveyor could be shrunk, thereby increasing sortation throughput.

But analytical software can do more than provide comparisons of boxes on conveyor lines, freight rates, or supplier performance. It can do more than just describe the current state of affairs at a factory or a distribution center. It can look forward into the future and recommend alternative courses of action. When analytical software recommends specific steps to take, it's called prescriptive. Because prescriptive software can model what-if scenarios, it can outline different ways a supply chain might handle a product shortage or the exit of a transportation carrier from the marketplace. It can determine trigger points that prompt specific actions.

Transportation is one area in the supply chain where prescriptive analytics are often used to recommend alternative courses of action, say, switching a load from one carrier to another. "Transportation [management software] has long had some prescriptive analytics because most were built on an optimizational foundation," Klappich said. "Systems were good at operational planning within the delivery time horizon but what we see now is more focus on tactical planning outside the execution time horizon."

Not only can software be prescriptive but it can also be predictive, forecasting potential problems in procurement, manufacturing, inventory forecasts, or distribution before they happen. Because the software detects an event that could pose a problem to a supply chain flow, it could propose a fix. Hence, predictive and prescriptive software often go hand in hand.

Say the software spotted a trend in the making such as an uptick in truckload shipments of product from a regional warehouse to a retailer. The pickup in shipments could be a harbinger of changing consumer demand in a region of the country. The software could make recommendations to deal with this anticipated demand change. It could propose a shift in replenishment from one region with minimal demand to the region where the trend line shows an uptick. Or it could propose that a factory switch production, making more of the desired product. By envisioning future spending patterns, the software could even identify potential sourcing problems before they occur.

One company that's doing some interesting work in applying predictive and prescriptive analytics to its supply chain is the multinational retailer Tesco PLC, headquartered in Cheshunt, Hertfordshire, England. Tesco has been a leader in the use of computer analytics to solve supply chain problems for years. Back in 1999, I remember writing an article on how Tesco did a computer simulation of its distribution operation to decide whether it should put up a standalone warehouse just for frozen foods.

In a presentation at the Teradata 2013 Partners Conference and Expo, Duncan Apthorp, head of supply chain development for Tesco, talked about a couple of analytical initiatives that the retailer had embarked on. One of the most interesting was its use of weather data to improve in-stock availability of fresh food such as meats and produce in its British stores, which number 3000. Tesco had discovered that when the temperature went up in the summer, it had a significant impact on shopper purchases. For instance, if the temperature rose 10°C, Tesco would sell 250% more barbecue meat, 45% more lettuce, 50% more coleslaw, and 25% fewer brussels sprout. The problem for the retailer was that its food suppliers need an advance notice of at least a week to make shipping adjustments.

To set up its analytical model using meteorological forecasts as the key driver, Tesco had to strip out other influencers of demand such as promotions, holidays, and day of the week (which impacts grocery sales). It then built a model for each of its products, correlating sales with weather based on past information. To forecast sales, the predictive model gets an automated feed with current meteorological information two to three times a day. The predictive model then generates a revised sales forecast that's shared daily with Tesco suppliers. The forecast is usually accurate for up to a week ahead, just a sufficient enough lead time for the supplier to ramp up. "Quite often we're making a call before we really know the weather is going to go the way we expect," said Apthorp at the Teradata conference.

The weather-driven sales forecasting has allowed Tesco to reduce out-of-stocks and have the correct item supply in the store when the weather swings. In addition, the retailer has saved considerable money by eliminating fresh-food waste that occurs when fair-weather consumers aren't in the mood for buying the meat or vegetables on hand at the store.

Tesco has also applied prescriptive analytics to store promotions. For the retailer, a large portion of its sales results from special offers. In the past, a buyer working with stock controllers at the store made the determination as to supply at local outlets to support a promotion. Since those stock forecasts were often rosy, the retailer applied analytics to the problem since the influence of discounts on shoppers can vary by item. A small discount will persuade a shopper to purchase meat reaching its expiration data. On the other hand, a deep discount is required to get the buyer to pick up a bag of salad. Based on its analysis of how discounts influence shopper purchases, Tesco has set up the model with price elasticity curves—the amount of discounts that will drive purchases by products. Every evening, store workers seek out products nearing their shelf life and apply new pricing labels with the right level of reductions to clear out inventory. Those prices are determined by the analytical software.

In the past, companies had to make a hefty capital investment to gain access to any type of business intelligence software. In fact, the expense was outside the budgets for most small- and medium-sized companies. A traditional business intelligence software solution was a often a seven-figure capital expenditure for the software licensing, computer hardware, and implementation, Steve Layne, chief operating officer for Trendset Information Services, stated in a February 2012 article in *DC Velocity*.

Due to the emergence of the "cloud" in computing software, business intelligence software has dropped in price. As the case with most other types of cloud software, the application can be rented online today as it's sold in a so-called software as a service (SaaS) model.

Companies no longer have to buy a software license and then install the application on corporate servers. Despite the emergence of cloud solutions, 85% of companies using business intelligence software back in 2012 were using on-premise application, said John Hagerty, a Gartner analyst at the time we spoke and now the program director for big data and analytics at IBM (TechWatch/*DC Velocity* magazine; February 2012).

Despite the availability of cloud software for analysis, data integration remains somewhat of an issue. Business intelligence software typically requires collecting information from a host of other applications, including enterprise resource planning system, warehouse management system, manufacturing execution systems, and TMS. The collected information was then stored in a repository known as a data warehouse. The central information depository had to be governed by master data management practices to ensure consistency. Without quality data, any analysis would be suspect.

One helpful development in recent years is the advent of data federation. Disparate databases located in different locations can be interconnected via a computer network. Data federation makes it possible for a business intelligence application to pull data from a source on demand. "It does away with the need to consolidate data from different source systems into another data store but it is not great for integrating large volumes of data or where there are big issues with data quality," said Sarah Burnett, an analyst with the Butler Group in the United Kingdom (TechWatch/*DC Velocity* magazine; September 2009).

Although data federation has drawbacks, it gives a way for the extended enterprise to have information visibility. To analyze an end-to-end supply chain operation that spans the globe and involves multiple suppliers, carriers, and partners, a company needs access to large amounts of information that may not be stored in a single central database or data warehouse. In many cases, the necessary information resides in an application that a supply chain partner is using for it alone—for example, information on the whereabouts of a truck and the temperature recordings inside the reefer trailer. A trucking company might have that type of telematics information in its computers, but it's not shared. Another example is information on what's being sold at the supermarket. And in some extreme cases, a partner may not hold the relevant information, although it's available on the Internet. For a demand-driven supply chain, even a remark on a blog or a social media site becomes an important piece of information as those comments could influence product purchases.

Vast quantities of information captured and stored somewhere that's relevant to an extended supply chain operation is referred to as "big data." In the world of information technology (IT), big data refers to volumes of information from multiple sources. Big data also connote a variety of sources. The data can be found in both structured information (residing in conventional databases) and unstructured information (residing in e-mail accounts or on social media websites). Finally, big data involve the speed of data creation. That's an important consideration for supply chains wanting real-time information to react and respond to unfolding events.

Information analysis of big data has captured the attention of companies as it can help them with all aspects of business from marketing to product development. And many businesses are already starting to mine big data for insights. In an interesting presentation at an Eyefortransport Big Data Summit in 2013, Steve Gold, an executive president at Opera Solutions, noted how his software company used big data analysis to help a retailer determine in what stores to use elastic or inelastic pricing. In certain locations, the retailer had the ability to raise prices (elasticity), and in other places, it lacked the ability to adjust prices (inelasticity) due to competition, whether from online merchants or nearby brick-and-mortar retailers. Big data analysis allowed the retailer to undertake the complex task of determining price elasticity down to the level of stock keeping units (SKUs).

But marketing isn't the only aspect of business well suited for big data analysis. The supply chain function is ripe for this. By its very nature, the operation of supply chain involves interaction between multiple parties and, in most cases, those parties not sharing all the relevant information with one another. Big data analysis could link all those relevant bits and pieces of information to create a big picture of the operation. "Think of pulling information from text— emails or comments—exchanged between people. Or may be comments left on a warehouse management system," said Aditya Naila, who worked for software vendor Dream Orbit, when we spoke. "Being unstructured data, traditional tools couldn't make much out of them and the data is obviously big. Logs generated by various software systems are huge and contain a wealth of information. Data might have been captured by traditional database systems but querying would be a challenge given the complexity."

For supply chain executives, big data analysis holds the promise of being able to find hidden connections and data patterns in different information repositories. "Software can identify patterns," said Kevin Sterneckert, a vice president of research in the consumer value chain at Gartner. "Typically, these techniques are applied to large data sets—aka 'big data'—where pattern detection is difficult. Social media, for example, is fraught with noise. However, buried within are valuable patterns that can be leveraged to help supply chains understand demand, plan the raw materials or components in production, and flow the merchandise through the supply chain to achieve 'on time and in full' [shipments of orders]."

Pattern analysis of data could result in the software raising both the question and the answer. In fact, Gartner analyst Noha Tohamy said big data analysis could lead to a fundamental shift in how questions are approached. "Right now we build a model to answer a question: where should I put inventory in my supply chain," she explained. "With a big data analysis you don't have to start with a question. It looks at patterns and correlations. It could find a relationship that might not be clear to the user. For example, poor customer service levels. Based on that correlation, what if we do 'x,' what will be the impact on the supply chain? It comes up with the recommendation."

Big data also expands the boundaries of conventional information analysis. As mentioned earlier, analytical software has been applied to a specific data set held in a software application to gain insights into warehousing, transportation, or production. But a supply chain is an extended enterprise, involving the coordination of many functions and departments within one company, as well as those of external partners. A supply chain has many touch points where partners hand off physical products and, in doing so, information to one another. A supply chain has flows where products and parts are moving and instruments are monitoring those movements in a factory, in a warehouse, or on a truck. So not only does the information required for a big data analysis come from multiple software applications but also the data must come from mobile and telematics devices, point-of-sale scanners, radio-frequency identification tags, smart sensors, and web-based platforms. As such, the data required to solve a supply chain problem are not likely to be stored in one place and in one company.

Only a few years ago, such a widespread analysis would not have been feasible. Big data analysis takes advantage of advances in computer hardware and software. Computers with bigger memories and faster processing speeds can sift through larger piles of data, whether in a structured format as found in a conventional database or in an unstructured format such as a blog posting. Big data analysis can be applied to both company-owned data repositories and external data warehouses of supply chain partners. And big data do not have to be texts or numbers. They can be pictures, sounds, or videos.

Big data analysis will also take advantage of developments in cognitive computing or artificial intelligence. Computers are being taught to think like humans. These thinking machines could result in new ways of seeing data, making connections, and solving problems. This field is sometimes called cognitive analytics. "Advances in cognitive analytic capabilities, machine driven learnings that get smarter over time . . . has the potential to further drive innovation and competitive advantage," wrote Richard Sharpe in his November 2013 blog, "How Can Big Data Drive Better Decisions?," for *DC Velocity*. "Cognitive analytics will be quickly adopted by those companies who have mastered the use of big data in their descriptive, prescriptive and predictive analysis activities."

The use of big data analysis could provide eye-popping operational insights for supply chain chiefs. Companies could learn how certain seemingly unrelated activities impact supply chain flows and processes. For example, supply chains could use big data analysis in solving shipping problems. Take "cold chain" shipping as an example. Pharmaceutical and food companies often transport products that must be kept at a specific temperature on their journey to the customer. Temperature sensors could be placed on the pallet for the ride in a truck with a cooling system. Those sensors could be then linked to another computer via a global positioning system. That way, software could monitor the temperature at given points in time during the truck's travels.

Here's how big data analysis comes into that picture. Those temperature history records could be cross-checked against the truck driver's electronic daily logbook. Since the logbook chronicles stops and starts of the vehicle, big data analysis could discover that certain activity such as vehicle vibrations in transit had impacted the integrity of a shipment of perishable fruits or vegetables, resulting in variations from a desired state of cold or warmth. Making the connection between multiple sets of data could not only improve the quality of the food sent to the store but also ensure food safety.

But the food supply chain isn't the only one that could benefit from sensor analytics. Any industry that deploys sensors could apply that data to its supply chain. Take airline industry as another example of where this might be useful. Sensor data on plane instruments could be shared back with the aircraft manufacturer as a way to improve preventative maintenance. The manufacturer could analyze the sensor data to determine how the aircraft components are functioning. That monitoring could let the aircraft manufacturer better predict when a part might fail on an airplane. Thus, the aircraft maker could prepare its supply chain for a quick response. Advance notification would allow the company to have available on hand the right parts to fix an instrument problem right away so the airline could get the plane back into the air.

Big data analysis can be used to ensure optimal supply chain flows. A path analysis would allow a supply chain manager to move a product more effectively from the point of manufacture to the customer. For this kind of analysis, data would be pulled in from a host of systems such as enterprise resourcing planning (ERP), materials requirements planning (MRP), warehouse management system (WMS), and TMS. This combined data pool could then be examined to get a holistic view of the flow. For example, a big picture analysis of the supply chain could pinpoint bottlenecks in manufacturing that impact distribution. "Information from a number of sources throughout the supply chain can help a manufacturer and their logistics partners understand exactly how product is flowing, where things are getting hung up, where value is added, or the location where damage or expenses occur," said Marilyn Craig, managing director at the firm InsightVoices.

The examination of large amounts of information, including nontraditional sources, could lead to the adoption of sophisticated "tipping point strategies" that would trigger certain actions. For example, a company detects that bunker fuel prices are starting to rise, thus jacking up the prices for ocean shipments. Since higher costs for ocean shipments undercut product margins, that could prompt the company to switch its production to another factory in a region of the world closer to the consuming market.

For companies engaged in a supply chain segmentation strategy, big data analysis can give them a more granular view of the "cost of serving" customers across the supply chain possibly in real time. Such an analysis could correlate segmented customer groups with cost-to-serve information found in such applications as transportation management or warehouse management systems. In his September 2013 blog for *DC Velocity* titled "Your Sales Are Costing Your Company Money," Descartes executive Chris Jones pointed that understanding "cost to serve" is perfect for big data analysis. In Jones' view, cost to serve "is a real opportunity for big data applications and one of the reasons you need a TMS [transportation management software] or fleet management solution to generate the granular data to determine where your company needs to change its customer service and order policy practices."

At present, a determination of the cost-to-serve analysis is generally based on a company's own internal financial data (see Chapter 5). But the use of external data could bring about a wider and clearer view of the cost-to-serve. Gartner analyst Tohamy told me about how an agrichemical company is already using big data analysis to develop different inventory policies for segmented customers. The company is factoring data about weather patterns and macroeconomics along with historical sales data to align inventory to customer product groupings.

Supply chain partners could use big data analysis for mutual benefit. Knowledge discovery in the extended supply chain takes on more importance when channel partners embrace a demand-driven strategy, as there's less room for error when it comes to replenishment. For example, a manufacturer could unknowingly cause problems for a retailer by using a trucking company that regularly damages freight in transit. Because the boxes containing the items arrive in decent shape, the retailer doesn't make the connection with the particular trucker doing delivery. A big data analysis, on the other hand, could correlate the problem with returned merchandise to a certain motor carrier employed by the manufacturer.

Big data analysis could prove fruitful in conducting a risk assessment of suppliers. For example, a company could use weather statistics on hurricanes, tornadoes, and floods to assess the probability that a supplier in a certain region of the world might be more prone to having its operations disrupted from a meteorological event. By identifying which suppliers are most at risk for bothersome weather, a manufacturer can devise a contingency plan for an alternative source or supplier in another region of the world, which would not be affected by similar climate conditions.

For companies with global supply chains, big data analysis could be used to link currency fluctuations with strategic sourcing decisions. Since the purchasing power of national currencies swing daily, changes in the exchange rate could make it more expensive or less expensive to buy materials, parts, or products in a particular country. By responding to shifts in the currency valuation, a company could swap suppliers, going from a supplier in a country with an unfavorable exchange rate to a supplier in another country where a favorable exchange rate boosts purchasing power.

Control towers are another area where big data analysis could be used to widen visibility and to improve coordination of supply chain flows. In the future, a control tower could gather financial data on suppliers and information on raw material price fluctuations and even particulars on geopolitical events. The tower could process all that disparate information to determine which supplier offers the best price for a component with the least amount of risk exposure from any kind of disruption. The tower could use that information to rank suppliers and have contingency plans in place to ensure a steady supply of materials. Most important of all, the control tower could do this analysis in real time.

Big data analysis could be used for retailers looking to adjust their supply chains to meet the challenges of e-commerce. Walmart already claims to be doing this. A spokesman for the company told me that for all of its online orders, the giant retailer is using big data analysis and automated intelligence to determine the optimal distribution to get the product efficiently to the customer based on various factors such as inventory availability, shipping costs, and timing. Walmart, by the way, is fulfilling its online orders from dedicated e-commerce fulfillment centers and store distribution centers with sections within them dedicated to e-commerce orders. Walmart also use its stores to ship online orders directly to customers.

Since extended supply chains involve many handoffs of parts and products with many actors, big data analysis can solve multidimensional problems. Although a company can analyze ways to improve transportation execution by poring through records of carrier performance from its TMS, that's just one dimension. To make transportation truly more efficient involves planning, another dimension. That requires data from customers and suppliers to get sharper insights into transportation planning. Specifically, in the case of planning, a company would need insight into future orders to determine future shipments to determine its need for carrier capacity—the availability of a truck on a certain route to move goods for store replenishment.

Big data analysis will involve getting access to data from supply chain partners. That could prove nettlesome as many companies want to keep their information under wraps. That stumbling block was called out in the 18th Annual 3PL study, which took a special look at big data in the supply chain. That 2013 edition of that study, which canvases both shippers and contract logistics providers, found a considerable degree of reluctance to information sharing. In that study, 32% of 3PL respondents and 22% of the shipper respondents said that they view their company data as proprietary. About 1400 companies took part in that year's 3PL study, which is sponsored by Penn State University, Capgemini Consulting, Korn Ferry International, and Penske Logistics.

Since partner data become critical to a holistic supply chain view, the dominant company in an extended supply chain may have to insist that partners share information. Indeed, the supply chain master of that extended enterprise may be forced to write into its contracts a requirement that suppliers, 3PL companies, and carriers furnish certain specified data.

Instead of the stick approach to sharing data, companies could use carrots. Incentives are one way a company could encourage its supply chain partners to share data, said Frode Huse Gjendem, who works in the consulting firm Accenture as its global lead operations analytics. Gjendem has worked on big data analysis supply chain projects for automobile and aircraft manufacturers.

One way to facilitate collaboration would be for the manufacturer to put in place at its expense the infrastructure required for data gathering at its suppliers. The manufacturer could then let the suppliers avail themselves of the analysis tools as an incentive for their cooperation. Should companies still remain wary of sharing confidential information, even if there's no cost to them, there's another step that could be taken to gain their agreement. Gjendem said a company could set up a neutral party to receive the information from supply chain partners. The trustee can keep the data from all parties confidential and take precautionary steps for data security.

But access to data is only one issue. Even if a partner is willing to share information, the data still have to be the right data. "The company has to understand what kind of data they have to process to get the right insight," said Gjendem. "You have to understand what kind of data you need. There has to be a data assessment before you can implement big data analytics."

Another obstacle to big data analysis in the supply chain may be the need for closer working relationships than now exists between IT department and the supply chain staff. The IT department may have to help obtain quality data for analysis. That point was sounded loud and clear in report in the 3PL study cited earlier. The report found that just 57% of shipper respondents in the survey and 47% of 3PLs said they had "access to timely and comprehensive data relating to supply chain planning and operations" within their organizations. Given the need for expertise in IT, data science and supply chains, companies may have to establish special teams to do big data analysis. Set up just for the supply chain, these teams would consist of both data scientists and supply chain experts.

Those obstacles notwithstanding, big data analysis remains compelling because it offers a way to gain crucial insights to find costs savings and new ways to make revenue. Because supply chains are complex with so many moving parts, conventional analysis of data won't work for protean supply chains. That's because information is not static. It changes constantly. "Data should be thought of as a flow to a reservoir," said Gold of Opera Solutions. "Data is constantly moving. That's why it's hard to get the right answer out of a data warehouse."

What protean supply chains need are not data warehouses but data streams—constant replenishments of information from a variety of sources into big data analysis applications. Those data streams will come from transactions in computer systems, sensor recordings, and even posts on social media websites. Since products and materials in a supply chain flow alongside information, devices monitoring their movements make it feasible for big data analysis in real time or, at least, close to real time. In fact, Sterneckert of Gartner said he expects big data pattern recognition to be used for demand sensing, demand shaping, and profitable response techniques.

The key here is the access to stream of up-to-the-second data. "You want to assess information as it's happening," Hagerty told me. "People think of analytics as data sitting around somewhere. It's about data in motion. It's about being able to make decisions in a far more rapid manner than if you look at things after they've been done."

Without data streams, supply chains will not be able to make operational assessments as events happen. As supply chains become demand-driven, they will have to be able to extract insights from the data coming at them from all directions and from dozens of places. In a presentation at CGT 2012 Business and Technology Leadership Conference, P&G demand planning executive Rafal Porzucek talked about how big data will have to come together with demand planning to give companies the ability to respond to fast-paced marketplaces.

As companies integrate retail and customer data into their demand planning, they will have the capability to produce what-if scenarios, using a variety of data streams. "We need to have that capability in combination with the big data," Porzucek said. "It will force much better, and much closer integration of work processes and systems and software design than what we see today. There will be no other way because we will not have more time to analyze more data. We will actually have less time to analyze more data."

Companies will have to evaluate those big-data-drawn, what-if scenarios through a financial lens to make sure that they make economic sense. "The analysis cannot be complete with the demand planning and volume forecast only," said Porzucek. "We will have to complement it with financial evaluation. In talking about the scenarios, and what-ifs, that volume and financial evaluation will have to come hand to hand ideally in real time. All of that will lead to work processes and systems that will combine demand, supply and business planning pretty much into one."

But it's not just one company that will have to do this all on its own by itself. Demand-driven supply chains require supply chain partners to synchronize procurement, production, and distribution, and that requires the ability to analyze data almost instantaneously for fact-driven course corrections. Thus, fast analysis of data becomes as important as big data analysis to protean supply chains.

BIBLIOGRAPHY

Duncan Apthorp. Supply chain optimisation through analytics. Video presentation. Teradata 2013 Partners Conference and Expo; 2013.

James A. Cooke. The case for business intelligence. *DC Velocity*. Posted online September 18, 2009.

James A. Cooke. BI moves to the cloud. *DC Velocity*. Published online January 17, 2012.

James A. Cooke. Getting insight from big data. *DC Velocity*. Published online July 18, 2012.

James A. Cooke. How Invacare slashed its freight spend. *DC Velocity*. Published online September 11, 2012.

James A. Cooke. A new way to spot problems in the DC. *DC Velocity*. Published online October 18, 2012.

James A. Cooke. Shippers show big interest in big data. *DC Velocity*. Published online July 15, 2013.

Thomas H. Davenport and Jill Dyche. Big data in big companies. International Institute for Analytics; May 2013.

Thomas H. Davenport and Jerry O'Dwyer. Tap into the power of analytics. *Supply Chain Quarterly*; Quarter 4, 2001.

Nigel Issa. Supply chain: Improving performance in pricing, planning, and sourcing. Opera Solutions white paper. 2013.

Chris Jones. Your sales folks are costing your company money! *DC Velocity* blog. Posted online September 16, 2013.

Mark Kremblewski and Rafal Porzucek. Video: Procter & Gamble: The road to rapid innovation for supply chain excellence demand driven data. Terra Technology newsroom; 2012.

Richard Sharpe. How can big data drive better decisions? *DC Velocity* blog. Posted online November 5, 2013.

CHAPTER 9

SHARED SUPPLY CHAINS

Building protean supply chains may not always be a solo effort. In some cases, companies may be better served working together. In dealing with rapidly changing market conditions, companies may be better able to jointly fashion a successful group response than to do an individual one. Certainly, working together allows companies to share the costs and the rewards from collaboration.

The idea of supply chain collaboration between trading partners is hardly new. There has been past industry attempts to encourage companies to work together to improve supply chain flows. Supporting collaboration between manufacturers and retailers became a key objective for the group, the Voluntary Interindustry Commerce Solutions (VICS) Association (now part of the standards group GS1). As discussed in a previous chapter, VICS championed a process called Collaborative Planning, Forecasting and Replenishment (CPFR) as a way for retailers and suppliers to work together on resupplying stores. CPFR was intended to lower overall inventories in the channel and to ensure that store shelves were full of the right product for the consumer.

Along with collective action on the part of industry organizations such as VICS, companies have taken steps on their own in the past

Protean Supply Chains: Ten Dynamics of Supply and Demand Alignment, First Edition.
James A. Cooke.
© 2014 John Wiley & Sons, Inc. Published 2014 by John Wiley & Sons, Inc.

to lower supply chain costs through sharing resources. Clifford F. Lynch, writing in his March 2012 column in *DC Velocity* magazine, noted that in the late 1950s, a public warehouse in Huntington, West Virginia, set up the first shared food consolidation program. Back then, most major food manufacturers used rail to ship customer orders rather than truck. Lynch wrote, "Utilizing one of the more bizarre rail tariff provisions in effect at the time, the public warehouse came up with a plan: It would consolidate compatible orders from multiple clients and load the merchandise into a separate rail car bound for each individual destination (up to three), thereby allowing its clients to ship goods at carload rates (plus stop-off charges) rather than at significantly higher less-than-truckload [LTL] rates." According to Lynch, that practice of coshipping became a hit with shippers.

Although there have been efforts in America for companies to work together on inventory replenishment and joint shipping, it's Europe to date that has witnessed the widest use of shared supply chains. Some manufacturers shipping products to the same retailer in Western Europe join together on warehousing and transportation delivery services. Not only is shared distribution more cost-efficient but it also means that the retailer has one truck rather than two or three trucks coming to its warehouse or store. By consolidating shipments for fewer deliveries, retailers can reduce their use of labor for receiving at their warehouses and stores.

One of the pioneers that forged shared supply chains in Europe was Kimberly-Clark, which, by way, is an American company headquartered in Dallas, Texas. Kimberly-Clark makes an assortment of well-known consumer products such as Huggies diapers, Scott's paper towels, and Kleenex facial tissues, to name but a few of their well-known brands.

In an article I wrote titled "Sharing Supply Chains for Mutual Gain" (*Supply Chain Quarterly*; Quarter 2, 2011) I detailed how Kimberly-Clark got involved with the practice of collaborative distribution. Back in 2003, some Dutch retailers who were early pioneers in basing store inventory on point-of-sale data wanted to replenish their store shelves on actual consumer takeaway. What they wanted Kimberly-Clark to do was make more frequent product deliveries to support demand-driven replenishment.

The problem for Kimberly-Clark was that more frequent deliveries would drive up its transportation costs. "The challenge for my customer was to find a way to reduce inventory in the supply chain without increasing our transportation costs," said Peter Surtees. At that time, he was Kimberly-Clark's director of supply chain for Europe when I did my interview for the *Quarterly* story. Surtees has since stepped down from this post, retiring from his position in 2013. He's now a director at the firm Collaborative Supply Chains Ltd. in Maidstone, Kent, in the United Kingdom.

To increase store replenishment, the personal care products manufacturer paired up with cosmetics maker Lever Fabergé (now part of the Unilever Home and Personal Care unit) to make joint deliveries. Kimberly-Clark and Lever Fabergé both filled up half the truck with their respective products.

Kimberly-Clark and Lever Fabergé did a trial with Makro, an operator of a chain of warehouse club stores in Holland. Surtees told me that the trial proved that collaborative distribution could lower store inventories while at the same time improving on-shelf product availability. As a result of that successful trial, Kimberly-Clark and Lever Fabergé hired a third-party logistics (3PL) operator to run a warehouse for housing their products and providing transportation on their behalf.

Kimberly-Clark went on to do shared supply chains with other companies in Europe. In 2007, the personal care products manufacturer started working with the multinational food provider, the Kellogg Co., in Great Britain. Both companies mixed their products to send full truckloads for delivery to stores. As was the case in Holland, 3PL companies handled the shared distribution in Great Britain for the two product makers.

In 2009, those two companies also began a similar initiative in France to serve French retailer Carrefour S.A. In France, however, Kimberly-Clark and Kellogg's each used their own contract logistics provider. Because the 3PL-operated facilities are near one another in Orleans, France, a motor carrier could easily make stops at both distribution centers to create a full truckload.

In the past decade, many other companies in Europe have begun initiatives to work together on distribution. One early example of this kind of cooperation was Project Sphinx, in which seven consumer packaged goods companies coordinated transportation shipments to six retailers in France. Since 2012, Project Sphinx participants have combined their products into full truckloads to make daily deliveries to stores. Benefits of the project include shorter replenishment cycles, enhanced customer services, improvements to on-shelf availability, and truck capacity utilization, Surtees said. Other benefits were reductions in retailer-held inventory, distribution costs, and greenhouse gas emissions.

Other collaboration efforts in France and Ireland were supported by a nonprofit organization formed to promote shared distribution in the European Community—the European Logistics Users, Providers and Enablers Group (ELUPEG). Formed in 2002, ELUPEG had more than 680 manufacturers, retailers, and logistics service providers as members of the organization as of 2013. The group trumpets that supply chain collaboration can reduce traffic congestion, lower carbon emissions, improve warehouse utilization, cut transport costs, and prevent trucks from running empty.

ELUPEG chairman Alan Waller said the group came into being as companies began to take advantage of the Single European Market, created in 1992, for the European Community and set up multinational supply chains in place of the previous national ones. As manufacturers and retailers started setting up cross-border supply chains, Waller said there was a realization that third-party logistic companies needed to expand beyond their national focus and take a more active role in setting up pan-European solutions. "The objective was to get people working together collaboratively to improve the performance of European supply chains," Waller said, "in particular, looking at asset utilization and bringing companies together to get volume economics. If you have the volume, you can get delivery frequency with good performance."

In an effort to further promote shared supply chains in Europe, in 2010, the European Commission provided 2 million euros in seed funding to form the group Collaborative Concepts for Co-modality (CO3). The mission of this consortium is to promote sustainable logistics through so-called horizontal collaboration, which is defined as cooperation across rather than along supply chains. By getting companies to comingle their shipments, the European Union (EU) aims to reduce Europe's dependence on foreign oil and to cut carbon emissions in transport 60% by the year 2050. It also wants to promote transportation modal shifts, moving shipments off trucks and onto rails or barges. The EU would like to move 30% of the trucks into another transport mode for shipments by 2030 and achieve 50% modal shift by 2050.

One of the main objectives of CO3 is to develop a legal framework for company collaboration as well as a solid business case for doing so. Because companies engaged in collaboration could possibly run afoul of antitrust laws and regulations, CO3 is developing a document that provides guidelines with a target date of 2014 for completion.

Key to the proposed legal framework would be the creation of a "neutral trustee" who would organize and manage companies involved in the shared supply chain. The use of a trustee means that the parties in the joint arrangement do not directly exchange information with another. Hence, only the trustee would be privy to shipper information about quantities, delivery dates, or specifications. "Legally, you need to put a firewall between competitors," said Sven Verstrepen in a presentation at the Council of Supply Chain Management Professionals (CSCMP) European Conference in 2013. Verstrepen is the business development director for a company called Tri-Vizor, which calls itself a supply chain orchestrator. Tri-Vizor is involved in a number of CO3 pilots on shared supply chains.

The trustee would not necessarily have to be a 3PL company, as has been the practice in Europe up until now. According to Verstrepen, the trustee could be a consulting firm or law firm as long as it's a neutral party. In fact, he told me that some 3PLs have become wary of multiparty collaboration as they worry about disclosure of individual pricing arrangements. They fear such disclosure could lead to a lowering of overall charges for their services.

Along with the legal framework, CO3 is also working on an operational framework and business model for companies to share costs and the gains derived from working together. "CO3 wants to do a business model to show companies how they can make money with collaboration," Verstrepen said.

As part of those efforts to prove the worth of the concept, the CO3 group has also been involved in setting up a number of pilots to demonstrate the value of supply chain collaboration. The organization looks for shippers with overlapping shipping routes and then approaches the companies about getting together on the transportation.

Health care company Baxter International, based in Deerfield, Illinois, is one of the companies Tri-Vizor has worked with on horizontal collaboration. It should be noted that Tri-Vizor began working with Baxter on coloaded movements before CO3 was formed. In one case, Baxter linked up with Belgian pharmaceutical manufacturer UCB to undertake joint shipments to six Eastern European countries in 2010. The two companies coloaded trucks to ship to Bulgaria, the Czech Republic, Hungary, Slovakia, and Slovenia. For the sixth country, Romania, the two coloaded a container for movement by rail.

In another case, Baxter teamed up with filtration system maker Donaldson to move raw materials from Belgium to Ireland in 2012. What's interesting about this collaboration is the use of a short-sea ferry service. After picking up the loads from Baxter and Donaldson, the truck takes a ferry to travel from the port of Zeebrugge in Belgium to Dublin, Ireland. From Dublin, the truck then goes to Castlebar, Ireland, where Baxter and Donaldson have plants near one another.

Companies involved in these cooperative arrangements place their shipment orders through a web portal set up by Tri-Vizor. The portal then assigns a carrier to move the load. The portal also keeps tabs on the shipments. It can provide data on the number of loads consolidated, pickup and delivery performance, and the monetary and greenhouse gas savings.

Ludovic Menedeme, director for transport and distribution services at Baxter World Trade S.A., told me that those two collaborations have resulted in double-digit freight savings for his company. He added that the company believes that it had lowered CO_2 emissions by as much as 30% from the consolidated truck shipments and the use of rail and short-sea shipping.

Another group in Europe promoting shared rail services is the Joint in Transport (JIT) Cooperative, which was formed in 2012 and supported by the Global Commerce Initiative, The Consumer Goods Forum, and consulting firm Capgemini. The aim of the Cooperative is to get companies to switch freight from the highway to rails and, in doing so, to curtail greenhouse gas emissions. In September 2013, the JIT Cooperative rolled out a first-of-its-kind intermodal service when three multinational companies—Colgate-Palmolive Company, Mondelez International Inc., and Nestle S.A.—moved shipments from Poland to the United Kingdom. The shippers trucked containers from their factories to a rail terminal in Poznan, Poland. From there, a train took the shipments to the port of Rotterdam in Holland, where they were placed aboard a ferry to cross the North Sea and reach Immingham, England. The Cooperative plans to increase collaborative intermodal shipments on the Poland-to-United Kingdom route and open up a south-to-north European route in 2014 for shippers in France, Spain, and Italy.

Although the concept of shared supply chains is under way in Europe, it's just getting going in the United States, even though many executives in American transportation and 3PL companies have argued that mutual distribution would yield enormous value. One American advocate is Chris Kane, chief customer strategy officer for the third-party provider Kane Is Able, based in Scranton, Pennsylvania. In his article titled "An Enlightened Approach to Distribution" (*Supply Chain Quarterly*; Quarter 3, 2010), Kane argued that if consumer packaged goods manufacturers would collocate their inventories, resulting in a shared approach to warehouse and trailer capacity, the net result would be a 35% reduction in product distribution. That reduction would occur because multiple sets of inventory would be consolidated into one regional distribution center where labor and storage costs could be shared. Cost reduction would also happen through consolidation into truckload shipments. Kane wrote that such an arrangement would also bring about "reduced chargeback fines as replenishment orders are delivered on an established scheduled agreed with the retailer." Chargebacks are fines levied by retailers when manufacturers don't adhere to specific shipping terms.

Kane went on to state in his article that supply chain executives needed to look beyond their own organizations to a wider universe of collaboration. "Moving to a collaborative model in which thousands of separate supply chains began to intertwine in a shared, intelligent, greener delivery network makes so much sense that it is a great deal more than a vision," Kane wrote. "It is an inevitability."

Kane's company is one of a few companies doing coloading in the United States. The 3PL Kane Is Able currently pools shipment for a number of mid-tier consumer products companies, including Sun-Maid and Topps. A Kane spokesman said shippers involved in coloading see the practice as a way of reducing freight costs and enabling retail customers to receive fewer but larger shipments of goods. Kane Is Able uses its own in-house transportation management software system to bundle loads to meet the retailer's requested arrival dates and delivery-time windows.

Since Kane wrote his article in 2010, interest has grown in shared supply chains in the United States, particularly in the area of transportation. At the 2013 Transplace Shipper Symposium, representatives from three companies—Colgate-Palmolive, Clorox, and Del Monte Foods—discussed their pilot program to coload truck shipments to a mutual customer. Facilitating the coloading project in the role of broker was Transplace in Frisco, Texas, a company that describes itself as both a technology and a 3PL provider.

In the past, all three companies had made their own individual deliveries on different schedules to the same customer. The program was designed to push up replenishment frequency while getting freight dollar savings from combined truckload shipments.

As of this writing, the initial coloading has proven so successful that Colgate-Palmolive, Clorox, and Del Monte were planning to expand the program and put on additional lanes. In my article "A TMS Is for Sharing" (*DC Velocity*; September 2013), Roger Sechler, director of transportation at Del Monte, said his company had witnessed freight savings from shipping truckload on a pro rata basis rather than less-than-truckload. "If you have LTL volume going to a customer and the other shipper is in the same geographical area, coloading makes a lot of sense," he said.

Because shippers can band together to consolidate their less-than-truckload shipments into a full truckload, coloading certainly helps shippers lower their overall transportation costs. That's a powerful economic incentive for shippers to consider adopting this practice.

But all the benefits of coloading don't just go to the shippers; receivers gain as well. As mentioned earlier, coloading means fewer deliveries that store or a distribution center has to accommodate. That results in less dock congestion for the receiver. And as noted previously, coloading can result in recurrent replenishment, allowing the retailer to maintain items in stock on the shelves at leaner inventory levels. Frequent, periodic deliveries reduce the need for buffer stock.

At the moment, Colgate-Palmolive, Clorox, and Del Monte Foods are pioneers for this type of cooperation in America. Although there are few others doing this, they tend to keep a low profile in the United States. But that could change because of pressure from certain large American retailers. If coloading gains wider adoption in the United States, the practice will likely be pushed upon consumer goods manufacturers by traditional brick-and-mortar retailers who seek to want to maintain in-store stocking levels as a way to ensure they have the right products on hand for a walk-in customer and don't further lose even more sales to web merchants.

The practice of coloading is easier to do if the shipping partners are using the same piece of software known as a transportation management system (TMS). That type of software is used for tendering loads to carriers and for scheduling pickups and deliveries. Certainly, having all the parties involved using the same TMS with identical features makes sense. (Different TMS packages could have different features.) In addition, and perhaps more important, the TMS can be set up such that the software automatically makes the shipper load pairings. The application also can automatically impose discipline regarding the terms of the coloading arrangement between the shipper parties; for example, a shared move will only occur if the shippers are going to the same customer within a 2-day time frame. "When you bring together freight that was managed separately, the TMS allows you to manage the requisite restraints to make sure that you make decisions that are executable," said Fabrizio Brasca, a vice president of industry strategy and global transportation at the JDA Software Group.

In the case of the Transplace pilot mentioned earlier, Transplace made modifications to its own TMS to facilitate the coloading test. Two of three shippers—Colgate-Palmolive and Del Monte—were 3PL customers so they were already using the Transplace TMS. As such those two companies had already fully integrated their processes and data to allow Transplace to handle shipment management. The third shipper—Clorox—submitted its shipment requests via electronic data interchange to Transplace. The Transplace TMS then merged those requests from all three shippers into a single consolidated load if a combination move met the needs of all the three parties.

Because Transplace is acting as the middleman, it facilitates the coloading process. That's not to say that coloading requires an intermediary. It's technologically feasible for two or more shippers to do the coloading themselves using a TMS. In fact, Brasca told me that one of JDA's clients is using a TMS to create consolidated shipments with a partner.

Although two shippers can in essence "share" a TMS to do coloading, it does raise of the issue of control. One shipper—likely the one running the TMS—has to be the dominant partner in the relationship and have final authority over the freight move. The parties involved can certainly write an agreement or contract setting forth the terms of the coloading arrangement as well as a mechanism for conflict resolution. And certainly, the TMS can assist in enforcement of the agreement terms. But since situations arise that are not planned for, one party has to be in charge. Relinquishing shipment control has been an obstacle to more companies teaming up.

Despite concerns over control, in the future, shippers may well want to give coloading consideration as a way to defend their delivery commitments to customers. A lot of industry pundits keep warning that trucking capacity in the United States is going to shrink in the coming decade due to a shortage of truck drivers. If shippers can't find a motor carrier to take a load, delivery service would be impacted. Sharing trucks would provide a way to address that looming problem.

Granted, many American shippers take a skeptical view of these warnings of driver shortages resulting in dwindling truck capacity.

They think the U.S. trucking industry has been crying wolf for years. Despite the dire warnings, a truck driver problem has yet to materialize. Keep in mind though, that the aftermath of the Great Recession has yet to see a return to full employment in the United States with other more lucrative job opportunities for truck drivers. Indeed, the economic downturn has kept many truckers on the job from retiring, as has been the case with many other lines of work where baby boomers don't have sufficient investment income to stop working. The lackluster economy has meant drivers can't find work in other related industries. Since many truckers don't like being away from home for extended periods as required in long-haul trucking, in good economic times, they often can find work in other industries such as construction that need individuals with a commercial driver's license for short-haul trucking. With the construction field in the doldrums, that has not been the case.

Although there are plenty of drivers today, a labor shortage of U.S. rig operators could very well happen in the next decade as the economy improves and current workers retire. Fewer drivers, of course, mean that truckers can put fewer pieces of equipment on the road. When trucking capacity issues do emerge, sharing supply chains would make it easier for companies to find a truck with a driver to get their products to the buyer.

Coloading, though, is only the transportation component of a shared supply chain. Warehousing would be another aspect. As discussed earlier, companies in Europe do that now, sharing warehousing space under an arrangement with a 3PL provider. One alternative to cowarehousing would be for companies to set up their distribution centers in an industrial park complex. That way, the distribution centers would be adjacent to one another to facilitate coloading.

Shared supply chains could be extended beyond warehousing and coloading. In theory, two or more companies could jointly run a trucking fleet to gain economies in delivery. Companies also could pool their orders for supplies in a joint purchasing arrangement to obtain a lower price in exchange for a volume commitment. They might even be able to go "halfsies" on the purchase of special software.

So, why is the United States behind Europe in setting up shared supply chains? I put that question to Peter Surtees, who now provides consulting services to business in regard to supply chain collaboration. His response was that scale and complexity of the U.S. market hasn't prompted as much need for collaboration. He noted that the scale of the U.S. market makes it more feasible for the manufacturer to send a full truckload of its own product to the retailer's distribution center. Because the cost of delivering smaller orders is considerably cheaper in the United States than in Europe, Surtees said American companies have less of a need to collaborate to achieve cost savings on freight movements. In addition, because Europe has smaller-sized stores than the United States, European companies want to reduce the order cycle time from manufacturer to store and more closely align replenishment with actual sales. Companies in Europe are also forced to collaborate as a way to mitigate the extra costs that result from having to make smaller but more frequent store deliveries.

I posed that same question as well to Professor Alan Waller. Like Surtees, he pointed to the difference in scale between Europe and the United States. He noted that cost-per-ton mile for delivery was higher in Europe than in the United States because shipments in Europe tend to travel shorter distances with smaller loads. But the creation of a single unified market also prompted a change in supply chain thinking in regard to collaboration. "Companies had run out of internal opportunities for savings. So, they looked externally," he said. "Supply chains in Europe are a little bit ahead of the curve."

I suspect that there are a few other reasons why shared supply chains gained traction first in Europe. Unlike the United States, Europe has fully embraced the environmental movement to combat climate change. The EU has set targets to cut greenhouse gas emissions by 2050 and transform that region of world into a so-called energy-efficient, low-carbon economy. As a result the EU and many western European governments are encouraging business and industry to institute programs to curtail greenhouse gas emissions. Collaborative transportation allows companies to combine product deliveries and, in doing so, to take trucks off the road shrinking their carbon footprint. Warehousing products under one roof rather than two does the same thing; economizing space means that energy and utilities can be reduced, again shrinking carbon footprint.

But since the United States has not adopted any formal national legislation to take action on climate change, there's no incentive for businesses to cooperate and collaborate on distribution on those environmental grounds, although some companies have started carbon-reduction programs in order to be good corporate citizens. Still, the push in Europe to "be green" has provided a huge incentive for sharing supply chains. In most presentations I've seen on the advantages of coloading, companies making and selling goods in

Europe always tout their joint efforts as their way to lend hands in the fight against global warming by supporting efforts to check the release of carbon dioxide into the atmosphere.

One of the biggest concerns in sharing supply chains is the specter of antitrust lawsuits. Companies that band together to restrict trade can be found liable for antitrust. To crack down on the abuses of manufacturing conglomerates, the United States in 1890 passed the Sherman Act, which was designed to restrict cartels and collusive acts on part of businesses to restrain trade. Although the Congress and courts have modified antitrust laws in the years following the passage of Sherman Act, it still remains a crime under many circumstances for U.S. companies to form a monopoly and raise prices on the consumer. In the 1980s, after the federal government deregulated transportation, shippers and carriers were warned that, at professional and industry gatherings, they were not to discuss any matter that might be construed as intended to thwart market competition. Although the U.S. government seems to have a more relaxed view of anticompetitive business activity at the moment, that stance could reverse with a change of presidential administration. Europe, on the other hand, seems to be interested in setting up a middle ground between anticompetitive behavior and cooperation. As noted earlier, the group CO3 is working on developing a legal framework for collaboration in Europe. Such a framework could spur further joint supply chain ventures and projects. If government regulators and industry can agree on a set of rules for sharing a supply chain, then companies in Europe could invest in collaboration without worrying that authorities might challenge those arrangements after an investment has been made.

Due to those antitrust concerns, companies in shared supply chains have often chosen to rely on the services of a contract logistics provider. By using a third-party, companies in collaborative distribution can argue that they are not engaged in any direct collusion to hinder competition. Since the 3PL company does the actual work of warehousing and transportation, the parties to the collaboration can make a "hands-off" claim. To my knowledge, this argument has never been tested in a court of law.

Besides the threat of antitrust action, there are other legal concerns about collaboration. For one thing, there are questions about legal liability when a company shares a truck with another company. There's also a question of control as discussed earlier. If two shippers wanted to combine loads without the services of a middleman like a 3PL, one company has to take charge. Any collaborative arrangement requires the involvement of legal counsels to write a contract setting out the duties and obligations of all parties.

Any contract would have to specify apportionment of cost and benefits, and that's another hurdle. How the gains are to be shared is an important question in the formation of any collaborative arrangement. Although in most cases two or more parties sharing warehousing or transportation should get a cost break, that might not always be the case. It's possible that a company might see its expenses go up on some types of shared arrangements. In that situation, companies would have to reallocate that individual company's increases in some manner to all parties. One solution, suggested by the group CO3, has been the use of Shapley value as the fairest way to divvy up rewards. Named in honor of Lloyd Shapley, the concept comes from game theory. But it's also a complex mathematical formula that, frankly, many just might not understand.

A shared supply chain requires a true partnership among the companies involved. And finding the right partner is another hurdle to joint ventures. "One of the barriers to collaboration is finding a partner you can trust," said Waller. "You need to trust the people you work with."

Another impediment to shared supply chains has been the combative mindset of chief executives. Many corporate bosses view their supply chain as a competitive weapon in the marketplace. The ability of a supply chain to provide superior customer service can provide differentiation when it comes to buying products, especially when the buyer views products as similar in quality and value. In a market where product prices are the same, service can be the tiebreaker, making a difference between a sale and no sale.

So if a company has figured out a way to deliver a product faster at a lower price, why share the secret? In fact, in 2010, the British conference organizer Eyefortransport did a survey of 580 logistics professionals on hurdles to horizontal collaboration. Shippers in the survey cited "fear of information disclosure" as the number one barrier. Frankly, that's not surprising. Most companies I've met don't want to let another business—even a nonrival—close enough to gain an understanding of any advantages possessed by its supply chain operations or, for that matter, any weakness.

The above-mentioned view holds true if a company sees itself as one against many in the marketplace. But in today's business environment, where suppliers can be tightly bound by to a specific brand owner or channel master, competition is actually supply chain against supply chain. Or put more precisely, its extended enterprise supply chain versus another extended enterprise supply chain.

By definition then, extended supply chains mean that there are multiple companies working in concert to bring to market a product that's ultimately delivered to a final consumer. Although some suppliers furnish raw materials, and others contribute parts to a final product, they all share a vested interest in boosting sales of a final product. More sales of the product mean more sales of materials and components.

It only makes sense that companies joined together in an extended enterprise supply chain would seek to coordinate their production and distribution to ensure a steady supply chain flow to the customer. It also makes sense that they strive to lower expenses on all of their behalf. If suppliers could share warehousing and transportation costs, then they can lower their collective bottom-line costs.

What's prevented this from happening? It's the mindset of corporate individualism. That's why the concept of "3-D supply chains" is so intriguing as it builds on this foundation of extended enterprise supply chains. Lawyer Ho-Hyung "Luke" Lee and Jess Parmer put forward the idea of "3-D" supply chains in a blog post on the Huffington Post website in 2012.

Lee went on to describe the concept of a 3-D supply chain in more detail in an article titled "How a 3-D Supply Chain Could Revolutionize Business" (*Supply Chain Quarterly*; Quarter 2, 2013). In Lee's view, supply chains started out initially as a "1-D network," connecting a company to its supplier and using either face-to-face communication or a telephone or telegraph to exchange information. Supply chains progressed to "2-D" models in which one company could exchange information with multiple entities. Think channel master and its multitude of suppliers.

Today, however, multiple companies taking advantage of the Internet can exchange information with many companies in a network, creating the basis for a 3-D supply chain.

Lee argues that these 3-D supply chains can be used to facilitate both competition and cooperation between trading partners.

How would this work? In his article, Lee uses the metaphor of a farmer's market to explain his idea. A local farmer's market entails cooperation in that growers must agree to a time, place, and rules for the venue of the market. Yet each grower competes in the arena of a farmer's market to make sales.

Lee envisions 3-D networks as a public hub unlike today's linear or 2-D supply chains, which are really private networks connecting suppliers, manufacturers, distributors, and retailers. He goes on to say this 3-D network would allow small- and medium-sized suppliers to compete against their larger brethren, even resulting in a reversal of offshored manufacturing and a return of jobs to the United States. I should note that Lee formed a company, Ubiquitous Market System (UBIMS), to provide an information-based supply chain infrastructure system.

Although I'm somewhat skeptical about Lee's idea of a public network, as it would require government support and regulation, his concept of 3-D supply chains remains intriguing. That's because it offers a way for companies to work together in a dynamic business environment. Certainly, information technology exists to provide a platform for multiple supply chain partners to connect to one another, to exchange information, and to facilitate trade. Industry trade groups could play a role here in promoting a hub.

Still, it may be unlikely for a 3-D supply chain to come into being in cases where a dominant retailer controls the channel and interface with the end customer. Nor will it happen in cases where the supply chain exists to support product innovation, say, a company that unveils new consumer electronics gadgets each year. But in markets where the products have become standardized or even commodities, it makes more sense for a group of companies, taking advantage of technology to facilitate mutual decision making, to work closely together and respond to changes as a group. Since they will need protean supply chains to respond appropriately, they need to be connected in an information hub to coordinate their responses to changing events.

Even if hubs don't materialize in the near future, shared supply chains will gain more adherents. That position was taken in a January 2011 report, "2020 Future Value Chain: Building Strategies for the New Decade," published by the group Consumer Goods Forum, an international organization of retailers and manufacturers. The report was the result of input from 200 executives representing retailers, manufacturers, third-party providers, and industry organizations as well as academicians. "Key trends such as increasing urbanization, consumer awareness about sustainability and the increased adoption of consumer and business technology will require new and collaborative distribution models and optimized 'smart' supply chain where information as well as assets are shared," said Consumer Goods Forum and Capgemini in their joint release on the report.

The report outlined a much broader vision for shared supply chains that any company is now practicing even in Europe. The report envisioned the collection of demand signals originated by consumers from home, in a store, on a stroll to the store with a smartphone. Those signals would be conveyed to manufacturers who would use them to make predictive forecasts on what to produce. The manufacturers would then ship their products to "collaborative warehouses" for "consolidated" deliveries to stores, pickup points, or a household. Not only would a shared supply chain lower greenhouse gases but the report asserted that it would also increase in-stock availability of products, improve order fill rates and customer service, and reduce lead times. At the same time, it would promote supply chain efficiency through cost reduction, fewer network nodes, and less handling.

Although shared supply chains have a considerable way to go to operate along the lines envisioned in the report, there are signs of movement in that direction. Waller told me that only 10–15% of opportunities for horizontal collaboration are being realized in Europe right now. Shared supply chains have also been confined to domestic movements in Europe and in the United States. But there's no reason to limit the geographic scope of shared supply chains. Coloading can certainly be applied to containers used for international shipping. Menedeme at Baxter, for example, has told me that his company would like to expand horizontal collaboration to include movements using ocean and air shipping between continents. Collaboration certainly makes sense for companies operating regional theater of supply. Suppliers located in an adjacent country to the buyer's home market could easily team up for comingled shipments.

The growing use of demand signals to drive supply chains is another impetus that will push companies to collaborate. As real-time demand becomes the driver for supply chain flow from production to consumption, it may no longer be cost-effective for companies to respond individually. Working together allows companies to gain some scale as a group in order to manage costs. Although companies are exploring ways to do this in transportation and warehousing, there's no reason why this concept can't be applied to manufacturing or even procurement. Certainly, companies could share factories or suppliers in theory. In fact, some companies already do share production resources. It's not uncommon for a contract manufacturer in China to make the same product for two or more name-brand companies. There's just no formal agreement about shared manufacturing.

As companies look to set up demand-driven supply chains, they will have to deploy control towers to manage the flows. As discussed in Chapter 7, these mechanisms can provide both visibility and response capabilities. The operation of a control tower allows partners in a supply chain to use information and then to synchronize their operations, making group decisions for proactive adjustments.

Although a control tower for a shared supply chain offers a way to control costs and to increase revenues, companies will have to overcome "mindset" issues on the part of chief executive officers and boards of directors to synchronized collaboration. After all, companies—not supply chains—are listed on the stock exchange. Corporate captains may have to get directly involved in formulating new agreements. Those agreements will have to spell out a "fair way" to allocate additional expenses and to ensure that members of the chain share in the expansion of the revenue pie.

Legal changes may also be required for tighter supply chain integration between companies. Although antitrust laws and regulations remain a nagging worry about supply chain sharing, at least in the United States although perhaps less so in western Europe, legal statutes may have to be amended or changed to keep current with the dynamic business landscape of the twenty-first century. If the law recognized that in today's global economy, that it's supply chains competing against supply chains for customers rather than company versus company, then that perspective could be applied to any legal scrutiny as whether an organization is engaging in monopoly conduct to push up prices outside of normal market forces. "The source of competition is in the branding and the offering," Waller told me. "Although supply chain has a competitive advantage in its totality, warehousing and transportation are commodities. You can get better leverage from working with people and you don't need to lose competitive advantage."

While waiting for governments to catch up and modify their laws, supply chain executives may have to take some risks. They may have to get out front and take a gamble. Supply chain chiefs may have little choice but to trailblaze new forms of collective action. Companies that work together in a dynamic fashion could well crush those that don't.

BIBLIOGRAPHY

Martin Arrand. Breaking the barriers to horizontal collaboration. Unipart Expert Practice white paper. 2011.

Companies to "test" shared supply chain in Europe. *Supply Chain Quarterly*; Quarter 1, 2011.

James A. Cooke. Sharing supply chains for mutual gain. *Supply Chain Quarterly*; Quarter 2, 2011.

James A. Cooke. Group developing legal framework for shared supply chains in Europe. *DC Velocity*. Published online May 20, 2013.

James A. Cooke. A TMS is for sharing. *DC Velocity*. Published online August 12, 2013.

James A. Cooke. It takes two (or more) for horizontal collaboration. *Supply Chain Quarterly*; Quarter 3, 2013.

James A. Cooke and Mark Solomon. Consumer packaged goods giants launch collaborative rail intermodal service in Europe. *DC Velocity*. Published online November 7, 2013.

European supply chain horizontal collaboration. Eyefortransport; January 2010.

Future supply chain in action: Multi-modal transport sharing initiative presentation. The Consumer Goods Forum and Capgemini; October 2010.

Horizontal collaboration in the health care supply chain. DHL white paper. May 2, 2012. Available at https://www.dhlsupplychainmatters.dhl.com/insights/whitepaper/106/horizontal-collaboration-in-healthcare.

Chris Kane. An enlightened approach to distribution. *Supply Chain Quarterly*; Quarter 3, 2010.

Ho-Hyung Lee. How a "3-D" supply chain process could revolutionize business. *Supply Chain Quarterly*; Quarter 2, 2013.

Clifford F. Lynch. A tried-and-true route to shaving freight expenses. *DC Velocity*. Published online March 20, 2012.

Maria Jesus Saenz. Horizontal collaboration and orchestration: How to save money and CO_2. Presentation Supply Chain World Europe; October 23, 2012.

Sven Schürer. Going a step further: Collaborating in the supply chain with a direct competitor. Eyefortransport conference presentation. Miebach Consulting; June 2010.

"3-D" supply chains could spur job creation, bloggers contend. *Supply Chain Quarterly*. Published online December 19, 2012.

2020 Future value chain: Building strategies for the new decade. Consumer Goods Forum, Capgemini, HP, and Microsoft; 2011.

Alan Waller. Supply chain collaboration past/present/future: The reality. Presentation at the Transport and Logistics 2013 Norsk Industri-ELUPEG Meeting; 2013.

Markus Zils and Carsten Wallmann. Identifying and assessing horizontal collaboration partnerships. Eyefortransport Conference presentation. McKinsey & Co.; June 1, 2010.

CHAPTER 10

A CUSTOMER OF ONE

One of the biggest challenges facing supply chain operations is the trend toward product personalization: serving a customer of one. Consumers and customers want it their way. Despite the best efforts of many marketing brands to build cache for their product, many customers consider themselves unique individuals. They don't want to fit a brand mold. They want a product designed to their personal tastes. Customers want to be at the center.

So, it's no wonder today that we find jeans makers setting up special shops inside the mall, offering to tailor a product in the store to the customer's own physique. A customer walks into the store, gets measured for his or her personal dimensions, and then gets a pair of jeans made that's fitted to his or her unique body shape.

In many ways, it's a throwback to what happened before the age of industrialization. A consumer walked into a tailor's shop to get a suit of clothes custom-made. A customer walked into a furniture maker's shop and got desk and chairs made especially for his or her house. A customer walked into a potter's shop and got a set up of cups and dinnerware.

But then large-scale manufacturing came along in the eighteenth century to end craftsmanship. Large-scale manufacturing could turn out products in volume such that the price of goods was affordable

Protean Supply Chains: Ten Dynamics of Supply and Demand Alignment, First Edition.
James A. Cooke.
© 2014 John Wiley & Sons, Inc. Published 2014 by John Wiley & Sons, Inc.

to the average person. Granted, a rich person could still visit a tailor to get a suit made, or visit a dressmaker for that special gown, but the average person bought his or her goods off the shelf or the rack at the store, merchandise mass-produced in size or style.

Although some manufacturers do engage in make-to-order production—Dell, for instance, configures computers to customer specifications—most product makers engage in make-to-stock production for consumer goods. And their supply chains support that approach to manufacturing.

The way products get made is soon about to change—thanks to technological developments that will result in a dramatic increase in make-to-order production. The technology that's expected to be the biggest driver for increased personalized production is additive layer manufacturing—better known by the name of three-dimensional (3-D) printing. Additive manufacturing holds out the promise of making it economical for a shop to produce one-of-a-kind items. In doing so, additive manufacturing could lay the basis for the Third Industrial Revolution.

3-D printers use a computer design as a virtual blueprint to make a 3-D object. Following that virtual blueprint, the printer deposits layer upon layer of material, fusing them together to make the object. This technology makes it possible to "click and print," producing a physical object from an image on a computer screen.

Additive manufacturing came into being in the 1980s. Notable pioneers were Charles Hull, who founded 3D Systems, and Scott Crump, who founded Stratasys. In the early days of 3-D printing, the technology was used to make product prototypes, as it allowed designers and engineers a way to touch and see their creations. Even today most sales of 3-D printers are still for that purpose as manufacturers use rapid prototyping to shorten the development life cycle for new products.

The advantage of 3-D printing is that this technology offers a faster way to make both parts and finished goods than current manufacturing methods. That's why it's a game-changing technology. Additive manufacturing changes the way products have been made since the start of the First Industrial Revolution when machines replaced handcrafted production. In traditional manufacturing, parts are forged, cut, and milled, and then they are assembled into a final product. Despite advances in technology, the manufacturing process is the still same one that Eli Whitney made famous in the early 1800s when he assembled interchangeable parts into a musket.

Since traditional manufacturing involves cutting, shaping, or grinding, it's a "subtractive" process in that material is removed or subtracted to make the desired form. A 3-D printer, however, adds layers together, hence the name additive manufacturing. The thinner the layers, the more exquisite the object produced. 3-D printers can produce some intricate geometric shapes that would bedevil sculptors to make.

For the virtual blueprint, 3-D printers can use data from computer-assisted design (CAD) packages, computer tomography (CT), and magnetic resonance imaging (MRI) scans. "The image is digitally sliced up into hundreds of two-dimensional layers, each representing a profile through the part to be manufactured," explained Phil Reeves in an article he wrote, "How Rapid Manufacturing Could Transform Supply Chains" (*Supply Chain Quarterly*; Quarter 4, 2008). "The layers are built up one at a time, from the bottom up, until the part is complete."

Not only can the top-of-the-line 3-D printers apply very thin sheets of material but they can also join different types of material. Layer manufacturing machines can fuse together plastics, ceramics, concrete, glass, and metal. Some 3-D printers can even use biological material such as tissue or amino acids. Indeed, pioneers in the field of medicine are experimenting with 3-D printers to make human body parts.

But 3-D printing could be used for more than just the manufacturing of physical goods and human organs. In the future, this technology could be used to make edible food. In 2013, NASA awarded a contact to the Systems and Materials Research Consultancy of Austin, Texas, to study whether 3-D printing could be used to make food as a way to feed astronauts on long space voyages. It may be possible at some point for 3-D printers to produce a variety of foodstuffs.

Although 3-D printing was first applied to making prototypes, its use for parts production has picked up in recent years. Back in 2003, final part production only represented 3.9% of the $3.9 billion spent on 3-D printing, according to the *Wohlers Report 2013*. In 2012, however, final part production had grown to 28.3% of the $2.2 spent on this technology. In fact, Tim Caffrey, senior consultant at Wohlers Associates, said he expects production of parts for final products to soon surpass prototyping applications for 3-D printed parts. "The money is in manufacturing, not prototyping," said Caffrey in a news release about the *2013 Wohlers Report*.

Holding back wider adoption of 3-D printers are its material and cost limitations. These machines generally can only use subsets within a material category. Take plastics, for example. Layer adding machines can only apply a few types of plastic for production at the moment. The cost of materials is also an issue. In its 2012 report, *3D Printing and the Future of Manufacturing*, the CSC Leading Edge Forum said the cost of feed material for 3-D printing ranged from $60 to $425 per kilogram compared with $2.40 for $3.30 for the same amount of material used in traditional injection molding. "Although the higher cost is not a problem for prototyping or small volumes, it is not economical for large volumes," the report stated. By the way, the Leading Edge Forum is a group within CSC that develops research, frameworks, models, and case studies for chief information officers (CIOs) and other senior business executives.

Even at the higher material costs, in some cases, 3-D manufacturing is already the preferable method for some substances. Take titanium metal. Titanium is hard to work with and even hardens during cutting, CSC Leading Edge Forum noted in its report. When titanium gets deposited in layers with a 3-D printer, that metal is more malleable.

Despite those existing material and cost restraints, manufacturers in certain areas have already adopted 3-D print technology to make a variety of parts used in product assembly. The *2013 Wohlers Report* said that final part production is being used for metal copings for dental crowns and bridges, orthopedic implants, and jewelry.

The automotive, aerospace, and medical industries have also begun using 3-D printing to make parts, said Sandeep Rana, a vice president of sales and operation at Additive Manufacturing Technologies (AMT) in Elk Grove Village, Illinois. One well-known manufacturer currently using additive manufacturing processes to make parts is the Seattle-based aircraft manufacturer, The Boeing Co. The company has produced more than 20,000 parts used on military and commercial vehicles in that manner. In fact, its new Dreamliner airplane contains 3-D-printed parts. Additive manufacturing at Boeing includes the following types: selective laser sintering, stereo lithography, fused deposition modeling, laminate object manufacturing, and electron beam melting.

Even though manufacturers today are using 3-D printers to either make components or prototypes, this technology could revolutionize how many finished goods are made. Depending on the size of the item, the 3-D printer could make the good on the spot. 3-D printers can replicate such items as dishware, hand tools, and sporting goods like rackets and balls.

Making a large-scale product such as an automobile or even a washing machine poses a challenge because of the physical dimension of the 3-D printer house. The printer itself would have to be bigger than the intended object of construction. To make large-scale products like an airplane with 3-D printers, manufacturers may switch their assembly approach from interchangeable parts to interlocking parts. Neil Gershenfeld, director of MIT's Center for Bits and Atoms, and Kenneth Cheung, a researcher at MIT, said they have developed a lightweight structure of composite pieces that could be put together to form an airplane fuselage. Gershenfeld said he came up with the concept in response to the question "Can you 3-D print an airplane?" Because he and Cheung deemed 3-D-printing a large object as impractical, they came up with a method using special composites that would facilitate the assembly of large structures from linking parts. Just as kids construct toys with Lego bricks, companies could build products with interlocked parts.

Whether manufacturers continue to use interchangeable parts or switch to interlocking parts, 3-D printing changes the factory process. To start with, it could eliminate the need for changeovers, converting a line or machine to make another product. Because of the cost of changeovers, manufacturing plants today are generally set up for long production runs of the same item. Because making a different thing with a 3-D printer only requires another computer design, the technology involves minimal setup time. Thus, 3-D printing could speed up the manufacturing of parts such that the factory could make the part only at the point in production when the component is needed for assembly.

If components can be made on the spot in the factory, there's no point for a manufacturer to carry any parts inventory on hand in the plant. Take a carmaker, for instance. 3-D printers could make a bumper at the exact point in the production cycle when the bumper is needed for assembly. The bumper can be made in the right size, style, and color for the automobile being produced. There's no need to keep and store an assortment of bumpers in a plant warehouse for just-in-time production.

Not only does additive manufacturing reduce or eliminate parts-on-hand inventory in the plant but it also does away with keeping inventories for after-sale product support. Why would a manufacturer want to make and store parts for out-of-date equipment when it could produce one on demand in response to a customer request? Instead of filling up stock in a warehouse, the manufacturer simply stores the designs for outmoded parts and products in a computer. In fact, equipment dealers or stores providing after-sales support could make replacement parts on the spot with 3-D printers. Even a repairperson coming to fix a household machine would not even have to carry spare parts. Instead of parts in the back of the van, there would be a 3-D printer.

Because the adoption of additive manufacturing requires less inventory, it could fundamentally transform supply chain operations and practices. Granted, the manufacturer would still have to keep some materials such as drums of powder, metal strips, or canisters in or near a factory for use by the 3-D printer. But the manufacturer could go from lean inventory to bare-bones inventory to make a finished product. In an article titled "Get Ready for the 'Software-Defined Supply Chain'" (*Supply Chain Quarterly*; Quarter 4, 2013), author Paul Brody, a consultant with IBM, described a 3-D printing test that demonstrated the need for less stock on hand. He said IBM Research conducted a test of making a hearing aid with a 3-D printer and determined that "shifting to 3D printing will eliminate more than 70 percent of the discrete components and materials—leading to a much simpler supply chain."

If companies using 3-D printing keep fewer parts or products on hand, that has a huge impact on space required for warehousing and manufacturing. Although the impact of additive manufacturing would certainly vary by product and geography, there would certainly be less need for companies to own or rent storage space. In a white paper report examining the impact of 3-D printing on supply chains, the global real estate firm Jones Lang LaSalle offered its views on how 3-D printing would change warehousing and manufacturing. The report, *The Evolution of Manufacturing*, said 3-D printing would "alter the type of premises companies require for manufacturing and blur the distinction between business, industrial and logistics space. In the future, for example, some warehouses could hold stocks to cover high demand items, but use 3D printers to print out low demand items. As a result these buildings would combine traditional storage and distribution functions with manufacturing." As for manufacturing facilities, the report said additive manufacturing would create the need for "more standard small and medium sized buildings, which companies would more likely lease than own."

But commercial real estate isn't the only industry that could feel a heavy impact from this new way of manufacturing. 3-D printing could well alter supplier and manufacturer relationships. If product manufacturers can make more of their own components, then they will need fewer outside suppliers. Unless the supplier can bring special expertise or knowledge to making a component, why would a durable goods manufacturer need them? Granted, it's not possible to make every component for a car or washing machine at this time with a 3-D printer. But a manufacturer will become more selective as to what parts they must obtain from a supplier and what parts they can make themselves. Finished goods makers will seek out strategic suppliers who can provide "special knowledge" and "capacity" to manufacture rather than the mere shipment of parts or components.

The suppliers retained by manufacturers will be situated near the plant as part of a company's regional theater of supply. As discussed in Chapter 3, companies will set up regional supply chains to serve key consumer markets around the world. Manufacturers in regional theaters will use primarily locally sourced components in conjunction with 3-D-made parts to ensure a quick speed-to-market for their finished goods. As more companies adopt demand-driven supply chains, 3-D printing will play an even more crucial role in facilitating speed-to-market production to reduce lead times.

Additive layer manufacturing thus revives local production to a greater degree than even reshoring for certain types of products and certain markets. As Reeves pointed out in his 2008 article, rapid manufacturing enabled by 3-D printers is an agile production strategy that "enables a truly distributed supply chain, where manufacturing can take place concurrently at multiple locations that are located close to the customer. This new structure could eliminate many stages of the traditional supply chain, affecting lead times, inventory management, and transaction and logistic costs."

More localized production with 3-D printing near the point of consumption reduces the need to ship parts and products long distances. The extended global supply chain will shrink and become a thing of the past in many industries. As manufacturers make more parts and finished goods in the plant with 3-D printers, they need fewer ocean, air, truck, rail, and barge shipments. Although transportation carriers will continue to move bulk commodities and raw materials, there will be an overall reduction in the use of transportation services for parts shipping.

Additive manufacturing will do more than just change the supply chains of current manufacturers. The technology will bring in new competitors who won't need much in the way of supply chain support. 3-D printing makes it possible for manufacturing to occur at the point of purchase. A customer could stop by a shop located in a village or mall, place an order, and then take the product home in one visit. In a digital world, a consumer could place his or her order online on an additive manufacturer's website, get the product designed to his or her desires, and then have the product delivered by an express delivery service to the residential door. In fact, entrepreneurs have already set up a couple of websites to do just that— 3-D-print small merchandise.

Just as important, the technology also makes it possible for manufacturing to occur at the time of need or use. When a consumer needed a product to do something—say, a wrench to unfasten a nut—he or she could make the product on the spot with a 3-D printer. In fact, there's even a website, Thingiverse.com, that offers designs for a home 3-D printer. The problem is that right now, the technology is outside the price range of the average consumer. Although a hobbyist can buy a 3-D printer for less than $1,000, production of sophisticated objects still requires an industrial quality printer. In fact, 3-D printers aimed at do-it-yourselfers mostly produce plastic objects unlike the industrial models. The *2013 Wohlers Report* said that a professional-grade industrial 3-D printer costs on average $74,340. Terry Wohlers, an expert on 3-D printing who puts out the *Wohlers Report*, told me that the lower cost 3-D printers lack the size, quality for resolution and feature detail, strength, speed, material variety, reliability, and ease of use found in the industrial models.

Still, if prices for high-end 3-D printers come down far enough and the technology becomes more consumer-friendly, it becomes practical for the average person to keep such a machine in their basement or garage for do-it-yourself production. Already hobbyists are using 3-D printers to make smartphone cases, coffee cups, utensils, and simple tools like wrenches. "A home-based manufacturer will need to stock small amounts of special materials at their sites and will likely be served by parcel and express carriers rather than by large transportation providers," wrote Reeves in the article cited earlier that discussed how rapid manufacturing enabled by 3-D printing would disrupt supply chains. "Just as iTunes disrupted the music industry and changed how music is purchased and delivered, the emergence of home manufacturing could collapse supply chains as RM [rapid manufacturing] transforms how certain types of products are purchased, produced and delivered."

Existing companies are already stepping forward to lend a hand to the do-it-yourselfers in getting started with making personalized products. In May 2013, Staples announced that it had begun selling the Cube(R)3D Printer that, in the company's words, was the "perfect printing device for designers, small businesses, students and kids."

A couple of months later, in July 2013, the UPS Store made an announcement that it was testing the concept of making a 3-D printer service available to customers. In the press release on the launch of a 3-D printing service, Michelle Van Slyke, vice president of marketing and small business solutions at the UPS Store, said, "Start-ups, entrepreneurs and small business owners may not have the capital to purchase a 3D printer on their own, but they may have a need to show prototypes to their current and potential customers. By offering 3D printing capabilities in-center, we're able to help further our small business customers' opportunities for success."

In the summer of 2013, The UPS Store launched the service in the San Diego, California, area with plans to add other U.S. locations. The company said the Statasys uPrint SE Plus printer being tested had the ability to print more detailed objects accurately than home 3-D printers. The printer could produce items like engineering parts, functional prototypes, acting props, architectural models, fixtures for cameras, lights, and cables. And, of course, once the item was made, the entrepreneur could ship the product direct from The UPS Store to the buyer.

The key point here is that for certain types of businesses and industries, 3-D printing could be one of the most disruptive technologies of the twenty-first century. Over the coming decades, people will be able to make more of their own consumer products rather than buy them, while established manufacturers of industrial products will face competition from upstarts for reasons to be discussed later in this chapter. Even though home manufacturers are not likely to make large durable appliances such as refrigerators and washing machines in the coming decade, the makers of durable goods will be forced to expand their use of additive layer manufacturing to produce parts in a bid to lower their production costs in a global economy. The result would be a further expansion of hybrid manufacturing process in which some of the assembled parts are cut and molded the old-fashioned way while other parts are produced with 3-D printers. In the long term, manufacturers could start constructing their products with interlocking components that can be put together to form an object just like plastic Lego bricks.

But 3-D printing will not just revolutionize manufacturing. It's going to disrupt the entire retail business model and the supply chains supporting the store channel even more than omnichannel commerce. Traditional brick-and-mortar retailers, along with consumer packaged goods manufacturers will face increasing competition from upstart 3-D printing shops and home manufacturers. Why drive to a store and hassle with traffic to pick up an item if it can be 3-D-printed at home? Why buy a name-brand product if there's an online design available for a generic equivalent? Why buy a replacement part for a broken-down piece of equipment if the part design can be downloaded from a website for reproduction on a 3-D printer?

To survive, retailers are going to have to engage in a thorough supply chain segmentation analysis to determine what the customer values and what the customer is willing to pay for. It will no longer be cost-effective for a brick-and-mortar retailer to carry every product on the store shelf, especially if it's competing against both web merchants and home manufacturers. Not only will the products on the shelf have to be those that the consumer will leave home for but also they will have to fetch profit margins. The fate of the retailer could well be determined by supply chain segmentation.

Gordon Fuller, a senior director at the firm CSC, said he believes that retailers are going to have to separate design from manufacture in determining what they offer in their store. If a consumer can download the design for a pair of eyeglasses, why buy them at the store? (By the way, Fuller told me that at this time, low-cost 3-D printers aren't available to use transparent plastic to make eyeglass lenses.) The retailer would have to offer products for which the consumer can't find an online design. In the eyewear example, the retailer would sell only lenses that can't be made at home, say, ones for anti-glare or UV ray protection. "Retailers will have to have sophisticated designs such that you still need to buy their products as opposed to a generic equivalent," Fuller said.

Retailers that compete on product innovation like Apple will have an edge in the altered retail landscape, said Fuller. Because Apple focuses on product design and outsources its manufacturing, its stores won't be affected as much by the home-production threat to retailing because the company still controls the intellectual property behind the product. On the other hand, retailers that don't offer unique or exclusive products will struggle. "If you're a transporter of other people's products, you're in trouble," Fuller said.

In the face of competition from home manufacturing, retailers may well have to respond by taking advantage of 3-D printing to become in-store manufacturers themselves. And that raises a lot of questions. What goods should a store stock on its shelves and what goods should be made in the store with a 3-D printer? Fuller raised a number of interesting questions related to this in the CSC report mentioned earlier. For example, should a retailer set up a special website to offer 3-D printer files for the customer to download? And if a customer makes the product, can the company offer any warranty or guarantee? And, finally, is the retailer liable for safety issues when it does not control the manufacturing?

It would not be surprising to see more retail outlets evolve into product showrooms with limited stock keeping unit (SKU) variety and limited on-the-shelf quantities of the product. Except for displays of seasonal bargain-priced products, the store shelf would only hold one item of a kind and that item would not even be there for sale. The shelf item would just be there for show. If the customer wanted a particular item, then the retailer would make the product on the spot with a 3-D printer.

As a result of additive manufacturing, the showrooming concept could spread across all segments of retailing to include retailers of large durable goods like appliances and automobiles. Take the car dealership as an example. A customer would walk into the automobile showroom, select the make, color, and style of vehicle with all its extra features, and then wait a few days while the factory assembles the vehicle specifically for the buyer. The auto factory would then ship the car to the dealer for customer pickup. Fuller said to me that in the future, carmakers might go one step further. The car manufacturer might ship the power train (chassis and engine) to the dealer and let the dealer do the product customization. The dealer would make the body parts with a 3-D printer to the consumer's preferences and assemble the car at the dealership. Thus, additive manufacturing makes possible personalized products even for durable goods. The color of the refrigerator can match the decor in the kitchen. The washing machine can be uniquely designed to fit into a cramped space of the laundry room.

As awareness grows about 3-D printing, consumers will make more demands on goods producers to make individualized products. If manufacturers don't comply, consumers will turn to another goods maker or do it themselves, where it's feasible. Despite the option, not every consumer will jump right away into do-it-yourself manufacturing. Frankly, I expect a higher rate of adoption of home manufacturing in rural and suburban areas of the United States. Handy folks in rural areas will embrace 3-D printing as it will allow them to avoid a lengthy trip to a distant store or the day's wait for delivery of an online order.

In urban areas, on the other hand, there will be less of a desire to own a 3-D printer if Amazon, Walmart, and other retailers perfect their capabilities to provide same-day deliveries for consumer orders placed online. Why have a high-end 3-D printer in a tidy apartment if a city dweller can expect a retailer to deliver goods within the time frame of an afternoon? For urban folks, it may come down to just plain convenience. It's a lot easier for a city dweller to buy a cookie from a bakeshop a short walk down the street from the apartment than go through the time, cost, and effort of baking. It will be interesting to see in the United States the extent to which suburbanites will make their own goods. The adoption rate of 3-D printing in suburban households may well determine the vacancy rate in U.S. shopping malls. And the shape of the retail landscape will determine the design of the supply chain network supporting it.

Although 3-D printing will upset the business model in discrete manufacturing (the making on an individual product) and retailing, the rollout of its impact will vary by market, product, and country. Western industrialized countries will be affected sooner by home manufacturing as more citizens there will have the discretionary income to acquire the technology. Although 3-D printing will eventually transform manufacturing and retailing, there are a number of short-term hurdles that must be overcome. In its white paper, Jones Lang LaSalle said that 3-D printing faces such short-term obstacles as "capital and operating costs, a restricted number of materials it can use, and the speed of printing."

Regulations may also produce barriers to wider technology adoption. For example, the U.S. government reportedly already has tried to stop the online dissemination of virtual designs to make guns with a 3-D printer. Trademark owners and manufacturers of branded wares may look to the courts to protect their intellectual property rights from copying by 3-D printers. Some consumer goods products such as pharmaceuticals may be less impacted by home manufacturing because of intellectual property rights and the lack of consumer confidence in the quality production of homemade medicine. Initially, do-it-yourself manufacturing may be confined to generic products, on which there are no existing patents, where the utilitarian value is just "good enough" for home use.

By the end of this decade, expect additive manufacturing to gain enough widespread adoption in the making of industrial and consumer products that supply chain operations and practices will be fundamentally altered. But 3-D printers are only of one of three emerging technologies that will upend business models and their supporting supply chain practices. In the article cited earlier, Brody argued that, in addition to 3-D printers, intelligent assembly robots and open source software for hardware designs will spell the demise of mass production. "The ability to manufacturer unique products in tiny volumes without having to worry about long-term spare-parts storage or long lead times is a revolution that is just now taking hold," Brody wrote in *Supply Chain Quarterly*.

The deployment of intelligent robots will play a role in this revolution as it results in faster product assembly at lower costs. In Brody's view, the price decline for intelligent assembly robots make automated factories economically feasible for small and upstart manufacturers. In the past, a robotic assembly station cost $250,000; now, those stations cost $25,000 and can be installed in a single day. A manufacturing entrepreneur would not have to hire dozens of assembly workers and instead could rely on robots.

The growth of open source software is the third element behind the dramatic change taking place in manufacturing. Designs for hundreds of products from mechanical systems to networking equipment are now available, according to Brody, who leads the electronics industry organization within IBM's Global Business Services unit. Open source means that technical specifications for making products will be free and available without copyright restrictions to any manufacturer.

Taken together, 3-D printing, robotics, and open source hardware designs knock down many of the barriers to competition. An upstart manufacturer does not need huge amounts of capital to purchase a 3-D printer or an intelligent robot. Open source provides access to product designs, intellectual property whose access in the past was restricted by patents. Because upstart manufacturers will focus on personalized products, the need for economy of scale for lower production costs is no longer in and of itself a barrier to market entry. Supplier agreements are no longer an obstacle as additive manufacturing will not require the need for contractors to make assembly parts. And all of that means century-old manufacturers will be competing with new entities to provide industrial products or the consumer products that home manufacturers are not making for themselves.

If that vision for the future of manufacturing is correct, then other factors will determine company success and profitability. In his article, Brody put forward the idea that the emphasis for product makers shifts from the manufacturing and assembly to supply chain "processes that can be defined, managed and executed through software."

Brody and his colleagues at IBM (which has done a white paper on this) termed this transformation in manufacturing "the software-defined" supply chain. The competitive differentiator then is the software management of the end-to-end process of bringing materials and components together to meet a specific customer need.

A software-defined supply chain certainly fits into the concept of protean supply chain discussed in this book. As described in the previous chapters, companies will have to create protean supply chains that can rapidly adapt procurement, manufacturing, and distribution process in response to constantly changing demands in global markets. 3-D printing, along with robotics, makes it possible for the manufacturing to be protean.

Although certainly many products with low economic value will still continue to be produced in batch runs for the foreseeable future, personalized production will increasingly become the major focus of manufacturing in the supply chain. Personalized production, by the way, is a strategy linked to all the others discussed in previous chapters in this book. Manufacturing supporting personalized production places factories closer to the point of consumption in regional theaters of supply. The design of the supply chain will have to be revised repeatedly as manufacturers reposition factories, likely small operations, and remaining warehouses, small facilities as well, to respond to the changing locus of demand. Companies will deploy control towers to have visibility to manage supply chain flows. Segmentation analysis will be used to assign inventory and service policies to different customer groups. Big data pattern analysis, along with prescriptive and predictive analytics, will be used to spot new demand trends quickly and to correct potential supply chain bottlenecks in the making. Companies will be forced to work cooperatively, competing as supply chains versus supply chains.

Personalized production reconnects manufacturing to the fundamental idea behind supply chain management: linking supply to demand. What the customer wants is what should be made. If true demand determines what is to be produced, then manufacturing must have the capability to produce to true demand. And the protean supply chain will change its shape in response to the demand. The only obstacle to this change will be sustainability, and that's what the next chapter is about.

BIBLIOGRAPHY

Paul Brody. The new software-defined supply chain preparing for the disruptive transformation of electronics design and manufacturing. IBM Institute for Business Value Executive Report. June 2013.

Paul Brody. Get set for the software-defined supply chain. *Supply Chain Quarterly*; Quarter 4, 2013.

Andrew Carleen. A new approach assembles big structures from small interlocking pieces. MIT news release; August 15, 2013.

Mark Cautela. Staples first major U.S. retailer to announce availability of 3D printers. Staples news release; May 3, 2013.

William Koff, Paul Gustafson, Jarrod Bassan, and Vivek Srinivasan. 3D printing and the future of manufacturing. CSC Leading Edge Forum report. Fall 2012.

James Morgan. Amaze project aims to take 3D printing into the metal age. *BBC News*; October 15, 2013.

Phil Reeves. How rapid manufacturing could transform supply chains. *Supply Chain Quarterly*; Quarter 4, 2008.

Alexandra Tornow and Jon Sleeman. The evolution of manufacturing. Jones Lang LaSalle white paper. June 2013.

The UPS store makes 3D printing accessible to start-ups and small business owners. News release. UPS Press Room; July 31, 2013.

The use of 3D printing for final part production continues impressive 10-year growth trend. News Release. Wohler Associates; November 18, 2013.

CHAPTER 11

THE SUSTAINABILITY PETARD

Many companies these days are paying fealty to the concept of sustainability. They have adopted corporate programs to support the mantra of "reduce, reuse, and recycle." A 2012 study by the Governance & Accountability Institute said that 57% of the Fortune 500 companies were publishing reports on their sustainability efforts. Although sustainability may make a business a good corporate citizen, sustainability does create some problems for protean supply chains.

The sustainability movement got under way three decades ago. A 1987 report by the United Nations World Commission on Environment and Development championed the idea of sustainable development. Sustainability was viewed as a way of meeting the needs of a growing worldwide population without having a negative impact on the environment. By the year 2000, many corporations had climbed aboard the sustainability bandwagon. Supply chains became a key focus of corporate sustainability initiatives because factories in the manufacture of products and distribution centers in the storage and handling use materials, water, and energy and create waste as by-products of their operations.

Protean Supply Chains: Ten Dynamics of Supply and Demand Alignment, First Edition.
James A. Cooke.
© 2014 John Wiley & Sons, Inc. Published 2014 by John Wiley & Sons, Inc.

Certainly it makes sense to use resources or energy more efficiently. Take electricity. A warehouse or factory can reduce electricity usage by taking advantage of daylighting in its buildings. Daylighting—using windows and skylights—replaces electric lighting fixtures with natural light from the sun. Advances in lighting technology have led to the creation of light-emitting diodes (LEDs) and compact fluorescent lights (CFLs), both of which use less electricity than incandescent bulbs. To keep a refrigerated warehouse both cool and energy-efficient, the building can use low-energy, solid-state LEDs in place of incandescent lighting.

Giant retailer Wal-Mart Stores Inc. has publicly talked about its efforts to implement energy-efficient lighting. In a 2011 presentation at a Dematic-sponsored conference on logistics and material handling, Jeff Smith, a senior director of logistics, maintenance, and purchasing for Walmart, said the retailer had replaced the lighting in more than 100 distribution centers over a period of 4 years. The lighting retrofit yielded an annual reduction of 1.9 million kilowatt-hours, resulting in an annual average savings of about $124,000 per facility, according to an article in *DC Velocity* magazine. If a company can install more energy-efficient lights in a warehouse or factory, and receive a quick payback, that's certainly smart business.

Along with retrofitting buildings for energy efficiency, another common focus of sustainability in supply chains has been recycling programs. In a 2008 story for *DC Velocity* magazine, I described how gourmet food maker Stonewall Kitchen in York, Maine, started a recycling program that separated waste components out of shipping material coming into its plant. Stonewall Kitchen in 2007 started its program by segmenting out cardboard and then later included aluminum and plastic. The company even earned some money by selling the recycled material. If a company can earn money by selling recyclables, that's smart business as well.

To get the most out of recycling, many companies have set up a reverse logistics program, bringing back discarded or used products to recover and reuse the material where possible. Many third-party logistics (3PL) companies specialize in helping companies with both reverse logistics and product life cycle management, particularly the later stages of recovery and disposal. Government regulations have played a role here in promoting this practice by pushing companies to manage the product life cycle within the supply chain. Electronic manufacturers, in particular, are required to take responsibility for their product returns. In a 2012 article in *DC Velocity* ("Handle with

Care"), Dale Rogers, a professor of logistics and supply chain management at Rutgers University, said that at least 27 of the 50 U.S. states had laws governing electronic waste.

Reducing packaging material and eliminating packaging waste is another focus of supply chain sustainability initiatives. If a company can cut back on the amount of packaging used in shipments, it can reduce waste, which often ends up going to landfills. One innovative way for companies to reduce waste is to rightsize their packaging to custom-fit the shape of a product. In March 2013, *DC Velocity* magazine reported how office supplier Staples was using specialized equipment that cut and creased boxes into the exact size required to hold an item in an order. By producing custom-sized boxes, the retailer expected to obtain a 20% reduction in the use of corrugated material and to see a 60% reduction in the use of air pillows, which are placed as a cushion to prevent item damage in shipping. Certainly, using less packaging material makes good business sense.

One corporate leader in this area has been the Hewlett-Packard Company (HP). The high-tech company eliminated more than 300,000 cubic feet of packaging material in Europe for laptop computer shipments by switching from wood pallets to plastic pallets. A 2008 article in the magazine *DC Velocity* ("The Green Team") described how HP got those savings. Because the plastic pallets were thinner, it allowed HP to pack more products on a skid. By the way, it was also able to reuse the plastic pallet several times. In another move to cut waste, HP changed the way it packaged ink jet cartridges, switching from a large clamshell packaging unit to a trifold cardboard package.

Those packaging reduction initiatives were part of HP's sustainability program. HP is among a number of well-known companies that have announced sustainability programs for their supply chain. Consumer products giant Unilever, for example, has said that it wants to double its revenue while halving its environmental footprint. When I heard Unilever's chief supply chain officer Pier Luigi Sigismondi talk at a gathering of supply chain professionals, he made the statement that none of Unilever's U.S. factories are sending any waste to landfills and that its electricity comes from renewable resources. "Sustainability is part of the core business model," he said at Council of Supply Chain Management Professionals' (CSCMP) European Conference in 2013. "Our products use less energy and water."

At the moment, European companies appear to be leading the charge for sustainability initiatives, a point underscored in an interesting 2013 survey by the firm AlixPartners. Of the 150 supply chain executives who participated in the survey, the AlixPartners' 2013 Executive Survey on Supply Chain Sustainability, 88% of respondents from Europe had corporate sustainability programs in place compared with only 68% of Americans. By the way, most of sustainability programs revolved around the use of recycled raw materials or truckload consolidation.

The biggest drawback so far to green initiatives in the supply chain has been the cost of those programs and questions about payback. Again, this point was underscored in the same AlixPartners' survey cited earlier. While 62% of European supply chain executives in that survey said they would be willing to invest in initiatives that do not produce positive financial returns, only 31% of Americans were willing to do so. And of those companies who are willing to spend money on sustainable technologies, nearly 60% wanted a payback within 18 months or less. Clearly, costs have been a barrier to wider adoption of sustainability programs in the supply chain.

But supply chain executives should not be wary about sustainable initiatives just on the basis of cost. Some initiatives may complicate efforts to run protean supply chains in a volatile business environment. Consider programs to reduce greenhouse gas emissions in the supply chain. Most environmentalists and many scientists contend that the rise in global temperature over the past century has resulted from the release of greenhouse gases such as carbon dioxide (CO_2) and methane (CH_4) due to human activity. Because of the reliance on fossil fuels for energy, manufacturing and transportation—two key supply chain activities—are seen as major industrial sources for the greenhouse gas emissions causing climate change.

So the thinking goes that if companies would replace their use of fossil fuels—gas, diesel, and coal—with alternative forms of energy such as solar, wind, hydro, or nuclear, then they would do their part in combating climate change. At a minimum, if companies were simply to use their fossil fuels more efficiently, that would help.

Before a company rushes into action and makes drastic changes to its forms of energy use, it should have a clear understanding of how its current energy usage in the supply chain causes greenhouse gas emissions. It has to know the amount of greenhouse gas emissions it's releasing and where it's releasing those gases in the supply chain. It has to know its "carbon footprint." "The term 'carbon footprint' refers to the total amount of carbon dioxide and other greenhouse gases emitted over the entire lifecycle of a product or service," wrote Georgina Grenon, Joseph Martha, and Martha Turner, all of the firm Booz Allen Hamilton, in an article titled "How Big Is Your Carbon Footprint?" (*Supply Chain Quarterly*; Quarter 4, 2007). "Typically expressed in tons or grams of CO_2, the carbon-footprint concept helps businesses and governments understand the relative amount of damage a particular product or service causes to the environment."

A measurement of the current carbon footprint can establish a benchmark for governments or companies seeking to set reduction targets for greenhouse gases. Concerns over global warming have prompted the European Union (EU) to set a goal of reducing greenhouse gas emissions in that bloc. Unlike the United States, 15 EU countries signed onto the Kyoto Protocol, a treaty negotiated in 1997 in Kyoto, Japan, that put forward national goals for industrial countries to limit greenhouse gases. In fact, the EU has settled on a 2020 target of cutting its emissions to 20% below what they were in 1990. Some member states within the EU have their own reduction targets.

Western European governments are pressuring companies to assist in greenhouse gas reduction efforts. Already one country in the EU, the United Kingdom, has begun requiring large publicly traded businesses to report their carbon footprints. As of October 2013, all U.K.-based companies listed on the Main Market of the London Stock Exchange must report their global greenhouse gas emissions. By the way, the Carbon Trust, a nonprofit, London-based consulting firm focused on developing a low-carbon economy, reported on its website that there is no prescribed methodology in the regulations for these reports, although the so-called Greenhouse Gas (GHG) Protocol Corporate Standard is considered the most well-established method.

Even though the United States has not passed federal legislation adopting similar carbon reduction goals for the nation, nor do financial regulators require publicly traded companies to issue greenhouse gas emission reports as of this writing, many U.S.-based multinational companies have started their own efforts at carbon reduction. Walmart has established a goal of eliminating 20 million metric tons of greenhouse gas emissions from its supply chain by the end of 2015 in large part through collaboration with its suppliers. HP in 2008 began publishing its aggregated greenhouse gas emissions for its supply chain and in 2013 set a reduction goal for itself and its suppliers. By the way, despite a lack of formal approval from

Congress for the Kyoto Protocol, federal agencies in the United States under the Obama administration have encouraged industry efforts to curb greenhouse gases. In that regard, the Environmental Protection Agency (EPA) has set up SmartWay program with transportation carriers as a way to promote efforts in this area. The EPA has begun regulating some types of carbon dioxide emissions after the U.S. Supreme Court ruled that the agency had authority to do so under the federal Clean Air Act. Adopted in 1970, the Clean Air Act was originally designed to restrict pollution and acid rain. But in a controversial decision in 2007, the highest U.S. court ruled that carbon dioxide and other greenhouse gases fell under the definition of air pollutants in the legislation. EPA rulemakings in this regard, primarily aimed at factory emissions so far, continue to meet with court challenges from some parts of industry.

While legal wrangling continues over the extent of federal authority to impose regulations regarding global warming, some U.S. companies, as noted earlier, are embracing efforts to combat climate change. In that regard, companies often set a goal for greenhouse gas reductions and then measure their progress toward that goal.

Before a company can set a target, it has to map its current carbon footprint to get an accurate fix on current emissions. Mapping breaks down the supply chain process (procurement, manufacture, and transportation) for each product and quantifies the gas emissions based on the form or forms of energy used. "Each product-specific chart should include all of the steps involved in the product's supply chain throughout its lifecycle, from raw-material production and distribution to manufacturing (including disposal of manufacturing wastes), finished products distribution, sale and use and disposal (including recycling)," wrote Grenon, Martha, and Turner in the aforementioned article. Because of the life cycle view, mapping can be a challenging exercise, to say the least.

At the moment, there are three common standards used in carbon footprint mapping, according to Massachusetts of Institute Technology (MIT) professor Edgar Blanco, who has done considerable research in this area. The first is the GHG Protocol, which was developed in partnership by the World Resources Institute and the World Business Council for Sustainable Development. Blanco told me that the GHG Protocol was designed for product measurements. Another standard is SCOPE 3, which affords an extended enterprise view of the supply chain. The third is the PAS-2050 from the British Standards Institute (BSI).

The Carbon Trust, which worked on the development of PAS-2050, has been a leader in formulating assessments for the amount of CO_2 used in bringing products to the shelf. The firm developed the Carbon Reduction Label, also known as the "black footprint label," which was issued in 2007. Although employment of the black footprint logo on the part of U.K. companies is voluntary, a number of well-known British companies such as Tesco, Walkers, and Kingsmill apply that label to some of their store products. While primarily a practice in the United Kingdom, usage of the black footprint mark has spread outside that nation as British companies start putting that label on their worldwide products.

Although the label originally was intended to give consumers an idea of the amount of greenhouse gas emitted in making and bringing a product to a store shelf, Carbon Trust spokesman Jamie Plotnek said business has made greater use of the label. In particular, procurement managers use the label to get an understanding of greenhouse gas emissions.

How hard is the process of making a carbon footprint for a product? Plotnek told me that the amount of time required to develop a black footprint label for a product depends on the number of suppliers and the complexity of the supply chain. "If it's a bread bag, which is polyethylene, that's made in one factory out of one material, it's fairly easy," he said. On the other hand, most carbon footprint calculations are more difficult especially since the data for the product life cycle must be gathered from the number of companies involved.

Another London-based group working with companies to address their greenhouse gas emissions in their supply chain is the CDP, formerly called the Carbon Disclosure Project. (The group changed its name as it broadened its sustainability work to include water usage.) In 2007, CDP began working with companies to get their suppliers to report on their greenhouse gas emissions and reduction programs. Dexter Galvin, CDP's head of supply chain, said it was Walmart who initially approached CDP about measuring the carbon footprint of suppliers.

CDP began its supply chain reporting with 20 member companies; as of this writing, it has 65 corporate members covering such sectors as retailing, banking, automotive, and consumer packaged goods. Each year, in April, CDP surveys the suppliers of the member companies in regard to greenhouse gas emissions in the supply chain. Member companies have CDP survey strategic suppliers, which usually account for 80% of the corporate purchasing expenditures, Galvin told me. In 2013, 2850 suppliers filled out survey responses, 450 more than the previous year.

Surveyed suppliers report on the greenhouse gas emissions in three areas. The first, termed "Scope 1," covers direct emissions such as those from the operation of a fleet of trucks. Scope 2 covers emissions from purchased energy, say, electricity used to power a factory. Scope 3 covers everything else and encompasses emissions across the entire supply chain from purchased goods through customer use. Scope 3 even includes business travel and employee commuting. Most suppliers use the GHG protocol as their method for determining carbon footprint. In addition to carbon footprint, the survey also asks suppliers about governance — who in the company is responsible for climate change reduction. It also asks about whether the supplier has set reduction targets.

The supplier responses then go into a report issued by CDP. Suppliers are rated on the completeness and accuracy of information submitted about their carbon footprint. Companies are also scored on whether they have taken any mitigation measures to reduce greenhouse gases. "We score all suppliers on transparency and willingness to provide information," said Galvin. "And we rate companies on performance, and the ambition of their reduction targets."

Companies such as Walmart and Unilever use the CDP's survey to track their suppliers' efforts to curb carbon emissions. Since many of the suppliers serve a number of the CDP company members, that means the supplier only has to fill out one survey. Those supplier reports are not just paper exercises. Galvin said that procurement departments of member companies are using the reports in supplier selection. "We have had members stop working with companies that were not willing to engage on this issue," Galvin said. "At this stage it's not about deselecting suppliers just yet. Our members want to work collaboratively with suppliers to help them improve."

Getting the information to fill out the survey can be time-consuming for suppliers that don't have software systems in place for data collection. In fact, Galvin conceded that when suppliers first fill out the survey, data collection can be "reasonably painful" for them. He added that "once you have the systems in place to get data, it's easy."

The exercise about collecting data does give companies surprising insights into the origin of greenhouse gases in the supply chain. Galvin told me that Walmart found that refrigerants used in its grocery stores accounted for a larger percentage of its greenhouse gas footprint than its truck fleet. By the way, a Walmart spokesman confirmed that story.

Indeed, it's not unusual for carbon mapping studies to yield some surprising results that run contrary to conventional thinking. For example, in 2008, PepsiCo's Tropicana beverage brought in experts to measure the carbon footprint of its orange juice supply chain. The experts measured the energy consumption at each stage of the life cycle and converted that consumption into the equivalent carbon dioxide emissions. The end result was an estimate of the greenhouse gas footprint for orange juice.

Now most folks would think that manufacturing or transportation was the biggest source of greenhouse gases. But that didn't turn out to be the case—the biggest source was the growing of oranges. Citrus groves use tons of nitrogen-based fertilizer and fertilizer production requires lots of natural gas. On top of that, fertilizer spread out on the field breaks down and converts into greenhouse gases.

One of the companies that have done pioneering work in carbon footprint mapping is yogurt maker Stonyfield Farm in Londonderry, New Hampshire. In 2001 and 2006, Stonyfield Farm did a complete audit of its carbon footprint that encompassed all company activities, including packaging and business travel. Milk production and packaging were the activities that released the most greenhouse gases. Cows release methane as part of the animal's digestive processes, and one molecule of methane is equal to 25 molecules of carbon dioxide, Nancy Hirshberg, vice president of natural resources at Stonyfield Farm, told me for an article I wrote in *Supply Chain Quarterly* in 2009 titled "On the Road to a Smaller Carbon Footprint."

The carbon mapping audit determined that transportation was right behind milk production in producing greenhouse gas emissions at Stonyfield Farm. The yogurt maker set up a program to devise ways to cut carbon dioxide emissions from its finished goods transportation in 2006. Its carbon footprint map encompassed Stonyfield Farm shipments from its main plant and distribution center in Londonderry as well as two copackers.

As a first step to reduce its carbon footprint, Stonyfield Farm consolidated its less-than-truckload shipments into full truckloads. To encourage its customers to order truckloads, it imposed order minimums and a 48-hour advance notice for order changes. Switching to truckloads allowed the company to eliminate four million miles of transportation moves, reducing its CO_2 per ton delivered by 40%.

The yogurt maker in 2009 also began using Railex, a weekly refrigerated railcar service for food products to move long-distance orders. The switch from motor carriers to rail was done because railroads generally emit fewer greenhouse gases than truckers. The company also upgraded its dedicated fleet of vehicles with newer units that get better fuel economy.

Although Stonyfield Farm was successful in shrinking its carbon footprint for transportation, it accomplished much of those goals through a reduction in transit miles as a result of shipment consolidation. It can be argued for the sake of efficiency, though, that any company should combine small shipments into bigger shipments if it wants to lower the expense of transportation. In fact, historically, many companies looking to cut freight costs have consolidated shipments into truckloads or switched to rail movements, which is generally a cheaper form of transportation than trucking.

Pairing shipping lanes is another way companies can reduce carbon footprint in transportation. *DC Velocity* magazine, in June 2013, did an article on two companies that did just that—Ocean Spray and Tropicana. According to the article, "Collaboration Bears Fruit," a 3PL company, Wheels Clipper, approached Ocean Spray about using a rail backhaul. One of the 3PL's clients, Tropicana, was shipping fruit by railroad from Florida to New Jersey, but the boxcars were returning back empty to the Sunshine State. Since Ocean Spray shipped a lot of products from Bordentown, New Jersey, the 3PL suggested that Ocean Spray use the rail backhaul. Ocean Spray's switch from truck to rail lowered transportation costs in the lane by 40%. That switch also reduced Ocean Spray's carbon footprint. According to an analysis done by MIT, the switch from truck to rail resulted in a savings of 1300 metric tons of carbon dioxide per lane for Ocean Spray. That switch also brought about an overall 20% emission reduction in Ocean Spray's distribution network.

To date, most carbon footprint reductions in the supply chain have been accomplished through traditional transportation efficiency practices. There's really nothing new about freight mileage reductions, changing the mix of transportation modes, running more energy-efficient vehicles, or increasing the value of density for shipped products. Dollar-conscious supply chain managers have been incorporating those practices for years.

The real way for carbon-lowering transportation would involve switching to trucks that use an alternative fuel source that generates less greenhouse gas emissions. Liquefied natural gas (LNG) or compressed natural gas (CNG) are the most talked about substitute fuels for trucking. Both LNG and CNG are considered "green" fuels because they release less carbon dioxide than diesel, the commonly used energy source for powering large trucks. Because CNG adds weight that must be carried along with the freight load, LNG is seen as the most practical alternative fuel for long-haul trucking—a distance of 400 miles or more.

Right now, buying an LNG-powered truck is a more expensive purchase than a rig running on diesel. In fact, a truck using LNG costs at least another $50,000 than similar equipment powered by diesel fuel. But LNG and CNG are attractive for U.S. shippers because the price of both fuels has been generally much lower than diesel in large part due to the abundant supply of natural gas within the borders of the United States. Indeed, LNG offers the added attraction of price stability for American shippers who have been hammered with uncontrollable freight rate spikes over the last decade as truckers impose fuel surcharges on shipping whenever the cost of diesel goes up at the pump. The biggest drawback to operating fleets of LNG trucks at this time in the United States is the lack of a nationwide network of refueling stations. LNG trucks must run on a defined looped delivery circuit to ensure access for refueling. In 2012, Dillon Transport Inc., a trucker based in Burr Ridge, Illinois, set up a program with Owens Corning running LNG trucks on fixed routes.

There would have to be a nationwide network of LNG refueling stations to support irregular route trucking. That infrastructure could well appear in the United States during the coming decade. That's because a number of big companies have started working with motor carriers to get them to use trucks running on natural gas. In June 2013, consumer products maker Procter & Gamble (P&G) announced plans to work with eight transportation carriers to convert up to 20% of its North American truckload shipments to natural gas-powered vehicles in a 2-year-time period. In its press release on those plans, P&G said the switch would reduce its greenhouse gas emissions by nearly 5000 metric tons.

Despite touting the potential carbon footprint savings, there's another big advantage to P&G in using LNG trucks. It's the hedge against volatility in the oil market. Because the United States possesses a vast supply of natural gas, even if a geopolitical event sends the price of oil soaring and raises the risk of diesel fuel scarcity, shippers like P&G would be able to obtain LNG at a reasonable price.

In the future, as natural gas takes hold in transportation, logistics managers would be wise to balance their shipping, using a mix with an equal number of diesel and LNG-powered trucks to maintain freight flows for a protean supply chain.

Although diversification of truck fuels makes sense as a risk mitigation strategy, some of the other strategies used to obtain lower greenhouse gas emissions come with a penalty. Keep in mind Stonyfield Farm had to get its customer to agree to a change in terms of service to use more full truckloads. In the case of Ocean Spray, it too had to adjust delivery schedules. Shipping by truck from New Jersey to Florida took 3 days, whereas the switch to rail meant a 4- or 5-day transit. To ensure supply, Ocean Spray's Florida distribution center had to carry additional inventory.

That's the rub for protean supply chains. In a world where customers are ordering personalized products online and expecting delivery the same day, there's a service expectation that results in more greenhouse gas emissions, not less. As consumers demand quicker delivery, there are fewer opportunities to gain transportation efficiency—and commensurate carbon dioxide reductions—through shipment consolidation or modal shifts. It's not practical to deliver online orders by railroad. Although couriers could use electric cars or parcel carriers deploy LNG truck fleets, there's still a question of economics. As companies grapple with higher costs for making same-day deliveries, can they afford to purchase higher-priced electric vehicles if the margins aren't there?

Carbon footprint reductions aren't the only sustainable initiative that runs counter to the creation of protean supply chains. Efforts to require suppliers to adopt sustainable practices could also complicate efforts for agility. Not only could sustainable practices raise the costs for suppliers, forcing the suppliers to raise their prices to maintain margins, but these initiatives could also have unintended negative consequences.

One of the biggest corporate champions for more sustainable supplier practices is Wal-Mart Stores Inc. Walmart got started with its sustainability program in 2005 after Hurricane Katrina struck the United States. In 2009, the retail giant began work on the "Sustainability Index," a measurement system for tracking the environmental impact of products. When Walmart launched that measurement system, its then chief executive officer (CEO) Mike Duke said the index would bring about a more transparent supply chain, drive innovation, and ultimately provide consumers the information they need to assess the sustainability of products.

The development of a sustainable standard was intended to be a 5-year project. In this regard, Walmart worked with the group the Sustainability Consortium to gather information for its Sustainability Index. The Consortium collects data around sustainability attributes such as greenhouse gas emissions, water usage, and waste for categories of consumer products. It develops category sustainability profiles for a product over its entire life. It also develops key performance indicators (KPIs), a set of questions that all the Sustainability Consortium members (not just Walmart) can use to assess and track the performance of brand manufacturers on sustainable issues.

A spokesman for the Sustainability Consortium said that its members use its data to find cost savings in their operations and supply chains through improved energy efficiency, lower use of raw materials, and reductions in waste. The Sustainability Consortium said Walmart uses its KPIs in the form of 15–20 questions for a given category to collect data from the retailer's suppliers about their sustainability practices as well as any product or supply chain innovations related to sustainability. For example, is the supplier measuring its greenhouse gas emissions, solid waste generation, or water usage? The supplier is also asked about whether it has established reduction targets in those three areas.

Walmart's buyers then use the Sustainability Index in evaluating supplier performance on sustainability. The index covers such areas as transportation, factory efficiency, and sourcing. As of 2013, Walmart said it had rolled the index across more than 200 product categories. Although Walmart had involved more than 1000 suppliers in those efforts to date, it was aiming to have 5000 of its suppliers on board by 2014. At this juncture, Walmart is working with suppliers to use more recycled materials in their products and to improve their energy efficiency. The giant retailer is also encouraging food suppliers to curb fertilizers in agriculture.

Although environmentalists laud corporations like Walmart for their work in getting suppliers to become more energy-efficient, to use less water, and to minimize waste, those programs could well backfire unless governments gets involved. If companies force their suppliers to engage in "green" practices that raise their costs, then a manufacturer or retailer could very well find itself paying more for its supplies than a competitor that does not follow green practices. That could handicap the supply chain operations of the green company, imposing additional costs and leading to higher product prices.

That's where government laws and regulations come into play to create an equal playing field for business competition. A federal law—assuming it would pass Congress—could force U.S. companies to buy goods only from suppliers who have certified programs to reduce waste disposal, curb water usage, or minimize their carbon footprint. A legal precedent exists in that the United States recently adopted the Conflict Minerals Law, which requires companies to show with an audit that they are not buying materials that were obtained through human rights abuses. Unfortunately, in a global economy, such federal laws could hurt American-based multinational companies, placing them at a competitive disadvantage against foreign rivals who are not subject to the same regulations. Indeed, the negative impact on U.S. manufacturing was one of the main arguments sounded in Congress against adopting the Kyoto Treaty. Congress saw no point in saddling American manufacturers with a "cap-and-trade" scheme in global marketplace if competitors in China and Russia did not have similar restraints.

Frankly, it's just as well Congress didn't pass a cap and trade for United States like the one that exists in many European countries. Emission trading is a poor scheme for curbing greenhouse gases as it does not address the underlying issue. Think about it. The government sets quotas or caps for companies on carbon emissions. Now if a company is below its quota, it can sell or trade its surplus rights to emit carbon to another company that's exceeding its cap. The only beneficiary is the government that charges for the permits.

Wouldn't a smarter approach to curbing greenhouse gas emissions be a tax on the supply chain? Such a tax would charge companies a set price for each ton of CO_2 emitted in manufacturing or transportation. Since business can be ruthless in ferreting out costs, most companies would quickly find any possible way to reduce or eliminate greenhouse gases to avoid paying a tax on their supply chain activities.

I raised that idea of a supply chain tax in a conversation back in 2009 with Danish economist Bjorn Lomborg, the author of the book *Cool It: The Skeptical Environmentalist's Guide to Global Warming*. By the way, Lomborg is quite critical of cap-and-trade schemes.

When I met him, he favored national governments promoting the development of alternative technology as the best approach to curbing greenhouse gas emissions.

Lomborg told me that any supply chain carbon tax would have to be set really high to goad companies into adopting greenhouse gas reduction programs that would raise their costs. But what really troubled Lomborg was that a supply chain carbon tax could wreck free trade. Nations with climate regulations could block importation of products from nations without such climate regulations. In short, a supply chain tax could be used as an excuse for protectionism and hence a barrier to trade.

The larger issue here is that there's no world policeman and there's not likely to be one to enforce supply chain carbon taxes, emission trading schemes, or corporate reductions in greenhouse gas emissions. In the absence of any enforcement or penalties, there's no practical reason for a company to raise its supply chain costs and to put itself at a competitive disadvantage vis-à-vis a market rival. Although a nation could penalize companies for failure to lower their greenhouse gas emission, it would jeopardize the ability of those national companies to compete in a global marketplace.

That's not to say a company should not engage in the husbandry of resources. For the sake of efficiency, companies should engage in resource management, whether it's oil, water, minerals, or energy. A company that uses fewer raw minerals in manufacturing lowers its production costs. A company that uses less water pays less of a water bill. A company that uses renewable energy to power equipment in its plant or warehouse can lower its electricity costs. A company that uses less diesel fuel or even natural gas to truck its products has a lower expense for transportation. So, resource management makes fiscal sense.

But sustainability issues don't make fiscal sense unless they can reduce costs. If a company consolidates less-than-truckload shipments into full truckloads and still meets its customer delivery commitments, then it makes fiscal sense to get the lower transportation rates that come with consolidation. But in cases, where sustainability raises costs, then there's no business case, except for good publicity. Despite all the clamor for sustainability, there's still little push from Wall Street for green. Investors remained focused on earnings per share.

"Most investors like a clean environment as well as the next guy," said analyst John G. Larkin, the managing director and head of transportation capital market research at Stifel, Nicholas & Company Inc. "However, they are less fanatical about it than are the hard-core environmentalists. What they are most interested in is value creation. If the green strategy doesn't reduce costs, enhance free cash flow, reduce asset intensity, make a product more strongly desired by customers, reduce inventories, and so on, then investors generally are less interested in whether a strategy is green or not."

Certainly some of the supply chain trends outlined in this book are favorable for both sustainability and profitability. As companies engage in nearshoring and set up factories and distribution centers closer to consuming markets, shipping distances are shortened, thus reducing carbon footprint transportation. Optimization of supply chain nodes—factories and warehouses—also shrinks shipping distances. The visibility afforded by control towers should also allow companies to plan shipments, leading to more optimized transportation moves.

But some trends may have the opposite effect, resulting in less sustainable supply chains. Take 3-D printing. Because it's an additive manufacturing process rather than subtractive, it should result in less material. But a closer examination of 3-D printing indicates that layer manufacturing just might not be sustainable at all. Research done at IBM, Paul Brody told me, modeled a variety of products, tearing them down, scanning the parts, and then rebuilding them using 3-D printers. Brody said he compared the carbon footprints of the 3-D printing solutions against traditional manufacturing and found that in some cases, additive manufacturing resulted in a higher carbon footprint. There were three reasons for that. Preparing feedstock for 3-D printers can be very energy intensive. Using 3-D printers is also quite energy intensive. And 3-D printing feedstock is not universally recyclable. "For some products we found, consequently, that the total carbon footprint was larger – between 10% and 20% – than traditional manufacturing (stamping, injection molding) even when netting out transportation benefits of more localized supply chains," Brody said.

Conceivably, arguments for sustainability could be used to stop the emergence of 3-D printing, which will become the dominant form of manufacturing for a protean chain. And that's the problem. It's not just that sustainable programs come with increased costs in some instances. The problem with sustainability is that it could impose new burdens at a time when supply chains must be protean to adapt to changing business conditions across the globe. The increasing push by leading companies to force their suppliers to reduce greenhouse gas emissions raises lots of thorny questions. Should a supplier be forced to change its manufacturing operation using a more expensive form of alternative energy that would lower its carbon footprint but raise its costs? What if the manufacturer's competitor did not and, as result, gained a price advantage in supply? How can the green manufacturer compete and still remain in business?

As can be seen, companies will be hard-pressed to meet the conflicting demands of sustainability and rising consumer demands that call for quicker supply chain deliveries. Consumers that want home delivery within an hour of an order placed online will force supply chains to engage in more frequent deliveries, not fewer. Although it's possible for companies to team up on transportation as discussed in Chapter 9, that's really only practical for line deliveries to the same grocer or retailer. How could a retailer engaged in home delivery team up to transport products to the individual's door? And if two retailers did team up and gained a price advantage against a competitor, wouldn't that be an antitrust violation?

Legislation for extended producer responsibility will also prove another burden, especially if companies are saddled with the legacy responsibility for discarded product lines and one-time-built products. If manufacturers must manage the entire life cycle of a product, they'll be required to set up a returns network to take back an out-of-use product and handle its disposal. Third-party providers specializing in reverse logistics will become essential partners for those companies operating protean supply chains.

Weighing the trade-offs between the costs for sustainability and operational impacts on the supply chain could very well become a required exercise for supply chain managers, especially those working for big corporations. In their article "Can You Be Green and Profitable?" (*Supply Chain Quarterly*; Quarter 3, 2008), Profit Point consultants Ted Schaefer and Alan Kosansky described a methodical approach to calculating the trade-offs of sustainability. They discuss the need to gather the right data, to set a baseline, to evaluate assumptions, and then to take a hard look at the costs for making structural changes. In this regard, big data analysis and other software tools for network evaluation will become necessary to see the big picture for total costs and to make data-driven decisions. Segmentation analysis will have to include reverse logistics.

But the chief concern isn't so much cost. It's the impact on responsiveness. Protean supply chains require a company to be able to change the makeup of its network in response to changing demands. Growing regulations for producer responsibility and carbon footprinting could hamstring supply chain flexibility. Since so many disruptive forces are sweeping retailing and manufacturing, I myself wonder whether some existing companies view regulations such as extended life cycle responsibility as a new way to create an economic barrier to entry for new competitors in a marketplace. If the government mandated that all companies only buy products from suppliers with a certain green score, compliance with extensive requirements for supplier certification could protect existing supply chains and their partners from upstarts who would have to spend considerable time meeting regulations rather than focusing their efforts on bringing new products or services to the market. The biggest complaint of small business owners is red tape as most lack the administrative resources of big business. Sustainability red tape could act as a counter to the nimbleness advantage entrepreneurs have in the marketplace.

But even the big companies promoting their own greenness might find themselves hoisted on the petard of sustainability. If a company has to calculate the impact of a carbon footprint whenever a change must be made to its supply chain, that will compromise its agility. Imagine having to take into account the carbon footprint whenever a supplier gets switched, a factory must be opened, or a new carrier must be hired. Imagine a supply chain forced to use more expensive manufacturing or transportation methods to comply with carbon reduction requirements. Although the software exists to do these carbon calculations, it could hamper decision making and lead to a hesitancy to act. For a demand-driven protean supply chain, indecision is the worst thing.

Sustainability could be the "fly in the ointment" for protean supply chains. Certainly, world population growth, along with increased life expectancy for the planet's inhabitants, will result in increased demands for energy, water, and resources. Although it makes sense for businesses to become more efficient in their production and distribution processes as a way to reduce resource consumption, supply chains must be able to shift their shape if companies are to survive and prosper.

BIBLIOGRAPHY

Peter Bradley. Collaboration bears fruit. *DC Velocity*; June 2012.

Peter Bradley. Handle with care. *DC Velocity*. Published online October 29, 2012.

Peter Bradley. That was easy (on the planet). *DC Velocity*. Published online February 26, 2013.

Carbon limits could harm trade. *Supply Chain Quarterly*; Quarter 2, 2009.

CDP supply chain report: A new era: Supplier management in the low-carbon economy. Carbon Disclosure Project. 2012.

James A. Cooke. Into the green. *DC Velocity*. Published online February 1, 2009.

James A. Cooke. On the road to a smaller carbon footprint. *Supply Chain Quarterly*; Quarter 4, 2009.

James A. Cooke. LNG: The road to stable truck rates? *DC Velocity*. Published online March 12, 2012.

James A. Cooke. All-star analyst. *Supply Chain Quarterly*; Quarter 4, 2012.

Foster Finley and Ron Scalzo. The 2013 executive survey of supply chain sustainability. AlixPartners; July 2013.

Pamela Gordon. New green strategies replace old notions. *Supply Chain Quarterly*; Quarter 4, 2009.

Georgina Grenon, Joseph Martha, and Martha Turner. How big is your carbon footprint? *Supply Chain Quarterly*; Quarter 4, 2007.

Group to develop standard measure for carbon footprints. *Supply Chain Quarterly*; Quarter 1, 2008.

Hewlett-Packard company announces supply chain greenhouse gas (ghg) emission reduction goals. Hewlett-Packard news release. September 23, 2013.

Martin Hickman. Carbon footprint labels: The latest aid for ethical shopping. *The Independent*; December 25, 2013.

How big is your carbon footprint? The U.K. wants to know. *Supply Chain Quarterly*; Quarter 3, 2007.

How much CO_2 is in your OJ? *Supply Chain Quarterly*; Quarter 1, 2009.

John R. Johnson. The green team. *DC Velocity*. Published online February 1, 2008.

Andrew Martin. How green is my orange juice? *New York Times*; January 21, 2009.

David Master and Lindsey Clark (primary authors). 2012 Corporate ESG sustainability/responsibility reporting: Does it matter? Governance and Accountability Institute Inc.; 2012.

P&G to convert 20 percent of its for-hire truck loads to natural gas. Procter & Gamble news release. June 27, 2013.

Brad Plumer. Supreme Court allows EPA to keep regulating carbon—But will review a few details. *Washington Post*; October 15, 2013.

Ted Schaefer and Alan Kosansky. Can you be green and profitable? *Supply Chain Quarterly*; Quarter 3, 2008.

Walmart, Grainger DCs achieve big energy savings. *DC Velocity*. Published online November 23, 2011.

Walmart highlights progress of sustainability index. Walmart corporate news release. September 12, 2013.

CHAPTER 12

FINANCIAL DYNAMISM

During the Great Recession (December 2007 to June 2009), more companies than in the past realized the connection between well-run supply chains and profitability. Clearly, chief executives have always understood that controlling costs in procurement, manufacturing, and distribution was critical to improving the bottom line. But in an anemic economy, there was a renewed sense of urgency to wring more cash from their operations, particularly the supply chain.

During the last few years, corporations have accumulated huge piles of cash and financial assets, what Bain & Company has called a "superabundance of capital." A global management company based in Boston, Bain has estimated that the total financial assets held by corporations exceeded $600 Trillion. And they expected that number to reach $900 trillion by 2020. That's gigantic! To put that number in perspective, bear in mind that in 2013, the U.S. economy—the nation's entire output of goods and services—was worth about $16 trillion. What's even more interesting is how much of those assets are plain old cash. Bain estimates the cash holdings of corporations extend beyond $1.8 trillion.

Not surprisingly, in response to the adverse economic headwinds of the last few years, corporations have become masters of cash

Protean Supply Chains: Ten Dynamics of Supply and Demand Alignment, First Edition.
James A. Cooke.
© 2014 John Wiley & Sons, Inc. Published 2014 by John Wiley & Sons, Inc.

liberation. A 2011 best practices report done by the group American Productivity Quality Center (APQC) and the consulting and audit firm Protiviti noted that companies—led by chief financial officers—became adept in the Great Recession of 2008–2009 in liberating cash "trapped in operations." The report pointed out that, by the middle of 2010, U.S. companies held close to $2 trillion in cash on their balance sheets. "CFOs [chief financial officers] will not stop harvesting unproductive cash from operations in the future—even when the skies are completely clear and it's time to aggressively ramp up growth investments," the report stated. "The effort to raise liquidity during the recession delivered some astonishing "lessons learned."

For corporate executives and boardrooms, the chief lesson from The Great Recession was this: Strong discipline in operations can release working capital that the corporation can use to raise profits which, in turn, boosts shareholder value. "Economic value to corporations is created by free cash flow—or the amount of revenues remaining after the funding of total enterprise expenses and taxes," wrote Gene R. Tyndall in the 2013 Tompkins International white paper, "Employing Available Capital Wisely."

Cash liberation in large part came from finding and freeing up money in the supply chain. Since working capital is the amount of dollars, euros, or yen used for a company's day-to-day operations, a more efficient supply chain saves the overall enterprise money. "Less capital employed typically translates into less money borrowed and less interest paid," John G. Larkin, managing director and head of transportation capital market research at Stifel, Nicholas & Company Inc., told me in an interview I did for *Supply Chain Quarterly* in 2012. "Less interest expense, in turn, enhances margins and free cash flows, either of which are often used by investors to value companies."

Along with freeing working capital during the Great Recession, companies focused on accelerating the "cash-to-cash" cycle time—the time period from when a product gets made to when it gets sold. For manufacturers selling products through retailers, or suppliers furnishing parts to an original equipment manufacturer, the wait time for getting paid became even more critical to survival in a tough economy. Without that cash, the company may not stay afloat.

Although the Great Recession may have officially ended, the world economy still languishes with growth tepid at best. Not surprisingly, the corporate mindset about cash liberation has taken hold and solidified as a way of doing business. In the New Normal economy characterized by sluggish business growth, the corporate obsession with working capital and cash-to-cash cycles is here to stay. As noted in the start of this book, companies have been enjoying record-high profits with stocks soaring on financial markets despite sputtering economies across the globe. What companies have discovered is that synchronization of supply chain components revs up the cash-to-cash cycle, the conversion of a product into revenue. In turn, those dollars, euros, or yen drive up shareholder value and for a public company, that's its raison d'être.

In his *Monetary Matters* column titled "What's Behind the Rise in U.S. Corporate Profits?" (*Supply Chain Quarterly*; Quarter 3, 2011), economist Chris Christopher, Jr. of the firm IHS Global described how supply chain efficiency, along with globalization and technology, raised profits during the Great Recession. "Currently, domestic corporate profits stand at 10.1 percent of GDP [gross domestic product]," Christopher wrote. "Analysts have been amazed at the rate at which domestic corporate profits have been able to spring back following the 'Great Recession' (December 2007 to June 2009). Indeed, domestic corporate profits as a percentage of GDP fell to their lowest point in over 60 years during the recession and then rebounded to their prerecession levels in just over one year. One reason why is that greater supply chain efficiency is allowing domestic companies to maintain lower inventories and avoid overshooting [accumulating inventory] when a slowdown occurs. This helps corporate profits to snap back more quickly. Inventory-to-sales ratios adjusted for inflation have been trending down for the retail and wholesale trade over the past 15 years."

Certainly cash liberation played a key role in generating those huge profits. Freeing up cash became an integral exercise for many publicly held companies during the Great Recession. One company that got a jump on that was Kraft Foods Inc., whose efforts in this area I wrote about in an article titled "At Kraft Cash Is King" (*Supply Chain Quarterly*; Quarter 1, 2010). (That company in 2012 split into two entities: Kraft Foods Group and Mondelez International.) It should be noted that Kraft Foods had made a corporate decision to free up cash to self-fund its business growth in 2007 before the economic downturn. It undertook a cash liberation initiative that had examined areas like payables, receivables, working capital, and capital expenditures. Because Kraft was running a complex, global supply chain, and individual business units had different inventory ownerships depending on the sales arrangement with customers, the company could not adopt a single enterprise-wide solution. It had to examine each business unit's supply chain separately.

Kraft tackled that challenge by creating its own internal team of experts to act as consultants for the business unit managers, who were offered incentives based on how much cash they freed up. Although each business unit settled on its own program for cash-flow liberation, a number of common tactics were employed. For instance, one tactic was rationalizing stock keeping units and phasing out low-revenue products with high demand volatility. Another was repetitive flexible manufacturing, in which the company broke up production into batches instead of manufacturing a type of product all at once. Another tactic was to reduce the fill rate on certain product orders. In regard to the latter, Philippe Lambote, who was Kraft's senior vice president of customer logistics in North America, told me, "When an SKU has a lower shelf velocity, it might not matter so much to provide high service since the customer's purchase frequency is not so high."

The goal of all those tactics was to reduce inventory holdings and hence free up cash. That supply chain initiative played a huge role in helping Kraft achieve its corporate objective of cash liberation. For the 12 months ending March 31, 2009, cash flow from Kraft's operation amounted to U.S. $4.3 billion compared with $3.6 billion in a previous 12-month time frame.

Although Kraft started its cash liberation initiatives prior to the onslaught of the Great Recession, its approach—looking for ways to free up cash—became a common strategy employed by business during the Great Recession. Because companies had to batten down the hatches and curtail expenses during the nadir of the economic downturn, they focused on efficiency. Supply chain synchronization—making procurement, production, and distribution work more closely in tandem—was one way to achieve savings and to free up working capital. Reducing inventory holdings offered the greatest opportunity and, as discussed earlier in this book, big companies in particular have become ruthless in their efforts to more tightly match stock with demand. Less inventory means less capital tied up. And quicker turns of inventory means acceleration in the cash-to-cash cycle.

Although the lean-and-mean mindset prevalent in corporate boardrooms will not change anytime soon, or at least until there's a worldwide upturn in economic growth, companies won't be able to merely streamline supply chain operations and practices to generate working capital in the coming decade as their strategy for profits. That's because the global business environment has intensified with burgeoning demands for personalized products and 24-hour-a-day service levels in the face of the onslaught of technological upheaval that will result in a third industrial revolution.

In many ways, companies played defense during the Great Recession. They took steps to cut logistics, manufacturing, and procurement costs. They took steps to cut inventory. They took steps to hold back payment to suppliers. That was all well and good as a corporate supply chain strategy when times first got tough in 2007. But the global marketplace has changed and new dynamics are at play as discussed in the previous chapters in this book.

The problem facing most companies and corporations is that the marketplace is about to become even more intense.

During the Great Recession, supply chain efficiency was aimed at improving the model of a static, structured supply chain. That approach is not going to cut it in a marketplace ruled by demanding customers wanting personalized products delivered minutes after being ordered. That approach certainly won't cut it for manufacturers and retailers if the customer decides to make the product with a 3-D printer rather than buy it. What's needed now is for companies to go on the offense. That means business has to look for ways to stay on top of the pace of marketplace change.

The world's largest online retailer, Amazon.com, Inc. represents an extreme case of a company in this regard. It's doing more than trying to remain abreast of change; it's trying to make supply chain change. In an article in *DC Velocity* magazine, the consulting firm Tompkins International said Amazon planned to expand its U.S. distribution network to more than 80 facilities by 2016. It's investing millions of dollars in a distribution network designed to provide same-day order fulfillment and delivery in metropolitan areas.

Although it's tough to know how much money Amazon is investing in supply chain and distribution infrastructure to provide online fulfillment, James A. Tompkins, the chief executive officer of the consulting firm Tompkins International, estimated that the e-commerce giant spend somewhere between $1.4 and $1.6 billion in 2013. "Perfect answers [on what Amazon is spending] are not clear due to Amazon accounting and what dollars they allocate to each year," said Tompkins.

Not only is Amazon setting the pace for service to online consumers but the giant web merchant has also begun competing against the likes of Grainger Industrial Supply for industrial buyers. It's pushed up the bar for service in e-commerce, whether for online orders placed by consumers or industrial buyers.

Certainly, retailers and industrial distributors will have to make some tough decisions as to how much money they are willing to invest in their supply chains to counter Amazon. But it's not just companies involved in e-commerce that are going to have to put some capital into creating protean supply chains. Companies in all business sectors will have to invest in giving their supply chains the capability to morph their shape quickly in response to changing business conditions such as the demand for new products and services as well as the emergence of new markets. In a dynamic marketplace, protean supply chains enable supply chain response that can increase and hopefully, if executed properly, restrain operational costs and minimize inventory such that there's a continued release of working capital.

In the past, change used to be incremental. A technology would be developed and often takes decades, if not centuries, to gain widespread adoption. It took ages for man to master the use of bronze, copper, and iron as technologies. What's different about today is that the pace of change has sped up. Even in the twentieth century, it took at least a decade or more for new technology to gain popular acceptance and usage. In modern history, it took decades for a majority of the population to own telephones, radios, refrigerators, stoves, and microwaves. But look what happened with the smartphone. Its adoption rate was at a faster clip than ever before. It has taken only a few years for the smartphone to be accepted by a large number of people.

Not only do humans accept technology quicker, the impact of adoption is far more widespread. Take the emergence of 3-D printers, discussed earlier in this book; it will go down in history as one of the most disruptive technologies ever invented. 3-D printing could fundamentally alter the way manufacturing and retailing has been practiced in the past century. That disruption is further complicated by heightened customer expectations for same-day delivery in the era of e-commerce. Given the need for dynamism, companies will have little choice but to deploy supply chains that can shift shape quickly. The supply chain must be protean to respond to changing conditions, to extend into new markets, to provide new services and products, and to serve the customer.

Make no mistake—this won't be easy. Chief executive officers need to recognize that the 10 trends outlined in the earlier chapters will impact their business and chief supply chain officers need to master those trends. Take forecasting. Historical forecasts are a thing of the past. Companies will have to base replenishment and production on actual customer or consumer purchases. Basing decisions on what to make and what to ship on demand will allow a company to maintain the leanest global inventory possible. Most of all, it's the best way to ensure an up-tempo in the cash-to-cash cycle time as stock will more tightly align to actual demand.

Setting up regional theaters of supply will enable manufacturers to use geography to their benefit. Cash liberation requires a speed-to-market approach to manufacturing and logistics. If you want to sell your product quickly and convert the product into the cash, then it makes imminent sense to make it in the region near or—better yet—where the consumer and customer are located. It makes imminent sense to source materials and components in a region near the customer, build the product near the point of sale or consumption, and then get the item into the hands of the buyer as quickly as possible.

Segmentation becomes an essential element of a supply chain strategy in the face of intense global competition. To maintain adequate profit margins, companies can't afford to spend money that might otherwise go to the bottom line. They have to divvy their customer base into segments and then align their supply chain service levels to each particular customer based on revenue and profit margin to the enterprise. Simply put, a business can't afford to waste money on unprofitable customers anymore. "If you segment by profitability, then you tailor your supply chain to the segment," said Jonathan L.S. Byrnes, a senior lecturer at the Massachusetts Institute of Technology. "Islands of profit [profitable customers] will be highly forecastable, flow through supply chains with very little safety stock and working capital goes down. For unprofitable customers, if they are not forecastable, you need a lot of working capital for safety stock in case they come in on one day and order from you."

For retailers engaged in omnichannel commerce, segmentation becomes absolutely critical to maintaining profit margins; in fact, without segmentation, most won't survive. Not only are web and brick-and-mortar merchants in competition with one another but they also will soon be in competition with home manufacturers. They will have to determine what products generate enough of a profit margin to buy from suppliers and to carry on the shelves. In a demand-driven supply chain, they will have to work with producers and suppliers to quickly make and replenish those shelves. As retailers and e-tailers start to taper their product assortment and restrict rather than expand stock keeping units, they will have to look over their shoulders and factor the threat of 3-D printing into every investment in technology or facilities for omnichannel commerce. "When looking at building new facilities, executives have to think how vulnerable their supply chain is to 3D printing," Fuller told me. "Right now it's 10 years away."

Decisions regarding investments in supply chain have to be data-driven. That's become more difficult since businesses are swamped with data and it's difficult to decide what information necessitates action. Retailers, manufacturers, and distributors will have to take advantage of analytical software to discern trends in the making to enable rapid supply chain responses to marketplace shifts. As soon as a trend is dictated, the company will have to revamp its supply chain. In some cases, that could entail building or shutting down plants or distribution centers. Analytics can judge the necessity and worth of such capital investments.

In many cases, the response to change requires a shift in the supply chain operation and that can be done with software, as discussed earlier in the chapter on segmentation. The warehouse management software in the distribution center can be programmed to assign a higher priority for inventory allotments to better paying customers and lower service levels to poor customers. Rules in transportation management software can be adjusted to allow for quicker delivery for some customers who contribute higher profit margins to the business and slower delivery to those who give little in revenue. As has been noted, software can be used to make supply chain adjustments in operation and not just physical assets.

Software that takes into account probability will become especially useful in proposing possible ways to address variances in demand. Such software uses a stochastic mathematical approach. And, in fact, some packages of supply chain planning and inventory optimization software already do use stochastic models in their calculations.

However, most inventory optimization and supply chain planning applications available today use a deterministic approach in their calculations. According to Tim Payne, a research vice president in Gartner's Supply Chain Core Research, a deterministic process means that given a particular input, the software will yield the same output or result every time.

On the other hand, when stochastic models are given an input, they produce a range of outputs. For example, if the service level required for the customer is 2-day delivery, a stochastic model will take into account that customer demand will vary over time into the software's calculations for the required stock level. Given the volatility of global markets and the randomness of demand, stochastic-based software is better suited to supply chain planning as these models more closely mirror the real world in which companies operate supply chains.

Lehigh University professor Larry Snyder described the stochastic approach this way. It "is basically an extension of classical optimization methods like linear programming in which the parameters are allowed to be random," explained Snyder, an associate professor of industrial and systems engineering and coauthor of the book *Fundamentals of Supply Chain Theory*. "The objective is usually to optimize the expected (average) cost or profit. Often the random parameters are described by scenarios. Each scenario specifies one possible 'state of the world.' Usually it is assumed that we know the probability that each scenario occurs."

Stochastic software can help companies prepare a plan for handling inventory, replenishment, or manufacturing orders based on the likelihood of certain events occurring. Stochastic optimization gives companies a way to prepare for random events and hedge against randomness. Snyder told me that he knew of a chemical company that used stochastic optimization models to plan inventory in the face of random blackouts imposed by the electricity company at the production site.

Instead of simply using a single number projection for inventory replenishment, as the case with deterministic software engines, Payne said that supply chain planning software using stochastic algorithms could set a range with upper and lower limits for determining stock replenishment. Inventory levels would be monitored to stay within a set tolerance. He has dubbed this restocking approach creating a "replenishment tunnel," a reference to the gap between the upper and lower limit. "These upper and lower limits are determined stochastically to allow for the correct buffering of demand and supply variability without sending nervousness back up the supply chain," he said.

As companies adopt a demand-driven approach to supply chains, stochastic planning may be become essential to steering supply chain flows. That's because stochastic software could develop ranges using demand data that mirror real-world fluctuations. It may be possible for stochastic models to use demand data to periodically recalculate the probability that certain supplies of products will be needed in the future to provide steady service to customers or buyers. The probability scenarios could be "refreshed" in response to actual demand. The software could keep adjusting the mix of options ranked on their likelihood of occurrence as demand for products or parts change. That way, the company could not get caught flat-footed if demand changes caused inventory shortfalls. Supply chain managers using this software could even tie their response into a segmentation strategy, ensuring that inventory gets set aside for the most profitable customer.

Although software based on stochastic models needs more development to achieve the vision described earlier, this application offers lots of potential. Not only will stochastic software point out scenarios to avoid excess inventory and save money but also, more importantly, they could point out possible opportunities for a company to earn revenue by better aligning supply with demand. But here's the catch. Supply chain executives may have to trust the software's output. Companies will have to let their inventory software set their safety stock automatically. As inventory software expert Shaun Snapp noted earlier in the book, companies have a tough time trusting the software's math.

To stay on top of a dynamic global market, companies will need to see what's happening in their supply chains in real time. If you cannot see it coming, you cannot react. Control towers give companies the ability to both react and respond to ongoing events as they happen. They enable protean supply chains by giving an organization the ability to find alternative sources of supply, to shift production, and to respond with alternative methods of replenishment. Without this ability to react, companies remain at the mercy of events. Slow reactions will result in lost revenue and higher expenses to fix problems at a later date. Control towers support a speed-to-market approach that becomes the best way to achieve cash liberation in a dynamic marketplace.

Not only must companies acquire enhanced visibility to alter procurement, manufacturing, and distribution but they also have to be able and ready to make modifications to the supply chain structure as business conditions dictate. Companies will need to frequently analyze the network layout and structure of their supply chains. They will require them to use software to model scenarios in anticipation of changing events and then make those changes to ensure supply chain synchronization and efficiency.

Although companies have gone it alone in the past, that too may have to change. In some cases, in the face of a difficult changing business environment, companies will be forced to band together and share their supply chains. Sharing will be one way to gain efficiency that could even free up working capital. Although sharing supply chain might entail a degree of legal risk, if governments decide to crack down on antitrust law, companies are going to have to weigh those risks against the economic benefits from cooperation and then explore the use of a trusteeship or middleman as a possible legal safeguard. In the end, business may have to lobby legislatures to make changes in laws to facilitate this sharing. After all, in the future, it's no longer company versus company. It's supply chain versus supply chain. For partners in a supply chain, their relationship becomes symbiotic with profits of each enterprise dependent on the others.

The drive for increased product personalization will prove both a boon and a challenge to business. Product makers embracing additive manufacturing will certainly be able to reduce their parts and components inventory, thereby freeing up working capital. On the other hand, additive layer manufacturing could wreck the structure of many supply chains. Durable goods and industrial manufacturers may eliminate many of their suppliers who cannot offer unique design expertise or special services. Home manufacturing will force both manufacturers and retailers to segment, concentrating on providing products that are unique or special enough to command adequate margins.

In an era of personalized production, companies have no other option besides segmentation strategies if they want to remain profitable. "In the prior 'age of mass markets,' companies sought the economies of scale of mass production, coupled with mass distribution using arm's length customer relationships," wrote Byrnes in a blog posted on *DC Velocity* titled "Profitability FAQ." "In that era, more revenues really did mean lower costs and more profits. In today's 'age of precision markets,' companies form different relationships with different sets of customers, each with different costs and profits."

Against this backdrop of steady fast-paced marketplace change, companies will confront the idealistic movement of sustainability that could well handicap them from being protean. If companies are forced to assess the sustainability impact of any proposed change to their supply chain, it will inhibit their ability to be competitive in the global marketplace. A lagging response in a dynamic business climate comes with a penalty. It could raise costs, but more importantly, it could cost companies sales.

The New Normal economy will be with us for many years to come. Lackluster business conditions in many countries, aging populations, the steady march of new, disruptive technology, and consumer demand for instant gratification will result in incessant turbulence in the marketplace. Constant change will require companies to be able to make prompt adjustments in supply chain structures and operations. Those that can morph their supply chains will be able to find success and reap profits. And those that can't will become dinosaurs. Expect many public companies to delisted or go out of business in the next decade.

Protean supply chains will enable the speedy response required for companies in the face of relentless change. "Time is the new measure of efficiency—time to market, time to benefit, time to delivery, time to cash, and even time to 'insight' are all measures of greater supply chain velocity," wrote Tyndall of Tompkins International in the white paper titled "Employing Available Capital Wisely."

In that paper, Tyndall argued that given the abundance of capital that now exists in company bank accounts, the time has come for businesses to use some of those funds and to invest in supply chain capabilities—technology, plants, fulfillment centers—to prepare for changes under way in the marketplace and to grow their business. "Because the cost of capital will remain low, access to capital is not the constraint facing companies," he wrote. "Rather, determining the best investment options will confound leaders for strategic capital deployment."

Without a doubt, companies should invest their capital wisely. Given the speed of change in the marketplace, to remain profitable, companies are going to be forced to embrace the thinking and practice of financial dynamism. In the past, companies could spend time developing a business strategy and then an operational or supply chain strategy to support the business objectives. That's no longer possible. Financial dynamism is the new mindset that matches the times. It should replace the lean mindset prevalent now in boardrooms. Financial dynamism means that when a company recognizes the opportunity for new revenue, it makes rapid investments in its supply chain operations. Although cash liberation and cash-to-cash cycle acceleration made sense at the start of the Great Recession, in the New Normal economy, it's financial dynamism that should guide executive thinking. A company has to have the ability to shift capital to new markets or to acquire new customers, to build new products, or to provide new levels of service or offer unique levels of service. In fact, the whole concept of the "order to cash" cycle becomes irrelevant if the product can be made instantly with a 3-D printer.

Financial dynamism goes hand in hand with protean supply chains. Only the ability to align products or supplies swiftly and precisely with demand will allow an organization to extract revenue at the highest possible margin. More importantly, this alignment will be the only way that companies can achieve sales and profit for their owners or stakeholders.

Although creating protean supply chain will not be easy, it offers a forward path. When companies recognize new revenue opportunities, or even the demise of existing profitable markets, they will have to alter or adjust their supply chain capabilities to meet those changes. "Supply chain capabilities cover the entire operation—people, process, locations and technology," said Tyndall. "Capabilities is the best word I know how to describe what you need operationally to deliver the business strategy. I got these products and that's why customers will buy from me in these markets. If your operations don't have the capability of delivering that, obviously you're not able to sell."

Companies will have to build organizations that can support rapid supply chain adjustments. To my knowledge, no company has done it. That said, many are working on aspects of improving their supply chain capabilities in response to the 10 trends in this book. But protean supply chains require companies to make modifications quickly. Although it's not feasible to shut down and open up a new factory or distribution center overnight, it is possible to modify software to change operational practices. It is possible to change the underlying business rules in the software program that determine which customer gets an allotment of product, which supplier gets tapped for components, which distribution center picks and ships the order, which carrier handles the delivery, and which product gets made in the factory.

In the end, public companies that don't embrace protean supply chains will not remain on the stock exchange. Private companies that don't adopt protean supply chains will cease to exist. And those that figure out the creation of protean supply chain will stay profitable for the years to come.

BIBLIOGRAPHY

Bain & Co. A world awash in money. November 14, 2012.

Jonathan Byrnes. Profitability FAQ. *DC Velocity* blog. Posted online December 5, 2011.

Chris Christopher, Jr. What's behind the rise in U.S. corporate profits? *Supply Chain Quarterly*; Quarter 3, 2011.

James A. Cooke. At Kraft, cash is king. *Supply Chain Quarterly*; Quarter 1, 2010.

James A. Cooke. All-star analyst. *Supply Chain Quarterly*; Quarter 4, 2012.

Improving working capital management and cash flow intelligence. APQC; 2011.

Tim Payne. Hype cycle for supply chain planning, 2013. Gartner; November 22, 2013.

Mark Solomon. Amazon: Just can't wait to be king. *DC Velocity*. Published online April 8, 2013.

Gene Tyndall. Employing available capital wisely. Tompkins International; October 2013.

Unchained mining the linkages between working capital and supply chain. Deloitte Development LLC; 2012.

INDEX

Protean Supply Chains: Ten Dynamics of Supply and Demand Alignment, First Edition.
James A. Cooke.
© 2014 John Wiley & Sons, Inc. Published 2014 by John Wiley & Sons, Inc.